Post-Kyoto Climate Governance

In the midst of human-induced global climate change, powerful industrialized nations and rapidly industrializing nations are still heavily dependent on fossil fuels. Even if we arrive at a Hubbert's peak for oil extraction in the twenty-first century, the availability of technologically recoverable coal and natural gas will mean that fossil fuels continue to be burned for many years to come, and our civilization will have to deal with the consequences far into the future. Climate change will not discriminate between rich and poor nations, and yet the UN-driven process of negotiating a global climate governance regime has hit serious roadblocks.

This book takes a trans-disciplinary perspective to identify the causes of failure in developing an international climate policy regime and lays out a roadmap for developing a post-Kyoto (post-2012) climate governance regime in the light of lessons learned from the Kyoto phase. Three critical policy analytical lenses are used to evaluate the inherent complexity of designing post-Kyoto climate policy: the politics of scale; the politics of ideology; and the politics of knowledge. The politics of scale lens focuses on the theme of temporal and spatial discounting observed in human societies and how it impacts the allocation of environmental commons and natural resources across space and time. The politics of ideology lens focuses on the themes of risk and uncertainty perception in complex, pluralistic human societies. The politics of knowledge lens focuses on the themes of knowledge and power dynamics in terms of governance and policy designs, such as marketization of climate governance observed in the Kyoto institutional regime.

Asim Zia is Assistant Professor in the Department of Community Development and Applied Economics and a Fellow at the Gund Institute of Ecological Economics at the University of Vermont, USA.

Routledge studies in ecological economics

Post-Kyoto Climate Governance

Confronting the politics of scale, ideology, and knowledge

Asim Zia

Routledge
Taylor & Francis Group

LONDON AND NEW YORK

First published 2013
by Routledge
2 Park Square, Milton Park, Abingdon, Oxon OX14 4RN

Simultaneously published in the USA and Canada
by Routledge
711 Third Avenue, New York, NY 10017

Routledge is an imprint of the Taylor & Francis Group, an informa business

British Library Cataloguing in Publication Data
A catalogue record for this book is available from the British Library

Library of Congress Cataloging in Publication Data
Zia, Asim.
Post-Kyoto climate governance : confronting the politics of scale, ideology, and knowledge / by Asim Zia.
    pages cm
    Includes bibliographical references and index.
    1. Environmental policy. 2. Climate change mitigation–Government policy. 3. Climatic changes–Political aspects. 4. Greenhouse gases–Political aspects. 5. Global temperature changes. 6. Nature–Effect of human beings on. 7. United Nations Framework Convention on Climate Change (1992). Protocols, etc., 1997 Dec. 11. I. Title.
    GE170.Z53 2013
    363.738'74561–dc23   2012032167

ISBN: 978-0-415-60125-2 (hbk)
ISBN: 978-0-203-06719-2 (ebk)

Typeset in Times New Roman
by Wearset Ltd, Boldon, Tyne and Wear

Printed and bound in the United States of America by Publishers Graphics, LLC on sustainably sourced paper.

Dedicated to Ursula, Anika and Rustum

Contents

Illustrations

Figures

Tables

Foreword

Just as the Belgian Congo gained its independence on June 30, 1960, the Belgian commander of the *Force Publique* was asked by his officers (all Belgians) about how independence might affect their leadership positions in the *Force*. He went to a blackboard and wrote, "After Independence Equals Before Independence." He was dead wrong. Four days later the Congolese army mutinied, Europeans fled the country and Congolese political leaders acted upon their newly gained sovereignty.

The same question can be asked in 2013 about a post-Kyoto Protocol world and how different it might be. My tendency is to paraphrase the commander's blackboard statement with a major difference: the post-Kyoto Protocol world will for the near term at least be similar to the Kyoto Protocol era. While the issues of concern may be better defined in the post-Kyoto world, the divisive stances of countries and groups of countries will remain with little sign of a desire for governments to change.

Robert Dahl, writing in 1961 about American politics, made a distinction between overlapping political cleavages in society and reinforcing ones. Cross-cutting political cleavages refers to the process of political bargaining: I will support you on issue A if you agree to support me on issue B. Antagonists realize that they may need the support of others they do not agree with in general at some time in the future. Overlapping cleavages refers to a situation in which the political groups tend to align themselves with likeminded groups on either side of an ideological divide thereby reinforcing their ideological differences rather than bypassing them. A good example of the latter would likely be the first four years of American President Obama's term in office where the two main political parties could not come to agreement on almost any political issue, a continually reinforced political divisiveness. As a result relatively little of his political agenda was accomplished in that first term in office.

In this treatise on his quest for a post-Kyoto global climate governance regime, Asim Zia is seeking a way to transition from the Kyoto world to a post-Kyoto world. He has identified the obstacles, the climate-related political cleavages: grandfathering the industrialized countries' emissions prior to 1990; a country's carbon dioxide emission levels vs. using per capita emissions levels to determine the equitable reduction in CO_2 emissions; binding cuts in emissions vs. non-binding cuts. The industrialized countries in general are on the same side of these issues while the developing countries stand together on the other side of the political cleavage. Zia's analysis seeks to find a path to an effective and equitable climate governance regime at the global level.

Technological solutions in the form of "geoengineering" are being proposed by the scientific communities in the industrialized countries—solutions that could in fact also threaten, not just this country or that one, this civilization or that one, but the habitability of the planet. In a 1957 article in *Tellus*, American scientists Roger Revelle and Hans Suess referred to the uncontrolled emissions of carbon dioxide as a global "geophysical experiment" the outcome of which was not known. They wrote the following:

> Thus human beings are now carrying out a large scale geophysical experiment of a kind that could not have happened in the past nor be reproduced in the future. Within a few centuries we are returning to the atmosphere and oceans the concentrated organic carbon stored in sedimentary rocks over hundreds of millions of years. This experiment, if adequately documented, may yield a far-reaching insight into the processes determining weather and climate....

Today, we know the outcome of uncontrolled greenhouse gas emissions—a heating up of the Earth's atmosphere—and those emissions continue to increase. Scientists do not believe that there will be a political solution to the urgent need for good governance of the planet's atmosphere. Hence, most "geoengineering" schemes seek a technical solution by "experimenting" with the Earth's climate regime in order to cool the Planet: millions of mirrors in space, mega-tons of iron particles in the ocean to increase algae blooms, mimicking volcanic eruptions by putting sulfates into the stratosphere, sequestering carbon deep below the Earth's surface, brightening clouds to reflect incoming solar radiation, mechanical carbon-absorbing trees, and the like. While there are many scientific "plans B" to try to cope with climate change, there is no "Planet B" if they fail. Are such schemes worth the risk?

International action on climate change (aka global warming) is needed now, but finding agreement among 200 or so participating nations in a short time is not realistic. Despite the appearance of extreme, even super, climate, water and weather events that are consistent with what researchers expect in a warmer Earth, those extremes are occurring not all at once. Perhaps a transition toward good governance involves at first the major GHG emitters, the G-20 countries, or the G-8 countries, or even the G-2, the United States and China. These two countries represent more than half the problem. In other words, what the world needs now is a C-20 (Carbon 20), or a C-8 or at least a C-2 (e.g., the US and China).

In the early 1970s top scientists came together in two workshops at the Massachusetts Institute of Technology (MIT) to address two major concerns: "Study of Critical Environmental Problems (SCEP)" and "Study of Man's Impact on Climate (SMIC)."

The SCEP report concluded that

- "Global problems do not necessarily need global solutions."
- "In the foreseeable future advanced industrial societies will probably have to carry the major burden of remedial action."

The SMIC report noted that

- "We recognize a real problem that a global temperature increase produced by man's injection of heat and CO_2 ... may lead to dramatic reduction even elimination of Arctic sea ice."
- "This exercise would be fruitless if we did not believe that society would be rational when faced with a set of decisions that could govern the future habitability of our planet."

One can only hope that today's world leaders will realize what is at risk for present and successive generations IF they fail to work together to bring about an abrupt, step-like change for climate governance in a post-Kyoto world.

I believe the creation of an EPCC (Engineering Panel for Climate Change) is called for; engineers gave us the good life with their inventions, so let's unleash them with the funds needed to engage in a "war on energy" to get us to a world whose life blood equivalent is truly "clean energy." Although wars on drugs or on prostitution or on poverty are not really winnable wars, a war on energy *is* winnable if only we can get politicians (all leaders are still first and foremost politicians) to be reasonable, to "satisfice" with regard to its growth and development as the right thing to do for humankind. In politics there is a "paradox of second best" in which second best options are acceptable to achieve one's goals, as in an election. However, for some problems there is no alternative to a known solution. Climate change is just such a problem. At the heart of the problem is the increasing emission of GHGs to the atmosphere as a result primarily of industrial processes in a divided political world. The solution is the need for clean sources of energy that do not enable the release of CO_2 into the atmosphere.

Michael H. Glantz
Boulder CO, USA
December 1, 2012

Acknowledgments

Countless individuals in many countries have helped me with shaping the ideas that are discussed in this book and I will not be able to list all of their names here. First of all, I owe a large thank you to the students in my graduate and undergraduate courses on Sustainable Development Policy, Policy Systems, Policy Analysis and Program Evaluation, and Decision-making Models here in the University of Vermont and a course on Global Climate Change at the San Jose State University for helping me with the continuous development of ideas and trans-disciplinary research presented here. Second, many governance and policy design issues presented in this book came to my attention from working on the following seven research grants during the last five years, which are listed in chronological order: (1) "Ecological boundary setting in mental and geophysical models" funded by the Human and Social Dynamics (HSD) Program of the National Science Foundation; (2) "A multiple criteria evaluation of greenhouse gas (GHG) emission reduction policies in the transportation sector of California" funded by the Mineta Transportation Institute; (3) "Advancing conservation in social context: working in a world of trade-offs" funded by the MacArthur Foundation; (4) "Navigating trade-offs in complex systems" funded by the U.S. Department of Transportation; (5) "Development of a complex governance systems research and modeling initiative" funded by the University of Vermont's Jeffords Center for Policy Research; (6) "Adaptation to climate change in the Lake Champlain Basin: new understanding through complex systems modeling" funded by the EPSCOR program of the National Science Foundation; and (7) "Understanding cross-scale dynamics in social ecological systems: integrative analysis of multi level institutional mechanisms governing tropical deforestation and forest degradation in Peru, Mozambique and Indonesia" funded by the Gund Institute of Ecological Economics. I remain thankful to my research collaborators on each one of these research project teams for bearing with me and helping me further develop these ideas. In addition, selected materials from the following four peer reviewed journal articles have seeped in various parts of this book and I am thankful to the respective publishers for giving me permission to re-use these published materials:

1 Zia, A. and A.M. Todd (2010) "Evaluating the effects of ideology on public understanding of climate change science: how to improve communication across ideological divides?" *Public Understanding of Science* 19(6): 743–761.
2 Zia, A. and C. Koliba (2011) "Accountable climate governance: dilemmas of performance management across complex governance networks." *Journal of Comparative Policy Analysis: Research and Practice* 13(5): 479–497.
3 Zia, A., Paul Hirsch, Alexander N. Songorwa, David R. Mutekanga, Sheila O'Connor, Thomas McShane, Pete Brosius, and Bryan Norton (2011) "Cross-scale value trade-offs in managing social-ecological systems: the politics of scale in Ruaha National Park, Tanzania." *Ecology and Society* 16(4): 7. Available at: http://dx.doi.org/10.5751/ES-04375–160407.
4 Zia, A. and M. Glantz (2012) "Risk zones: comparative lesson drawing and policy learning from flood insurance programs." *Journal of Comparative Policy Analysis: Research and Practice* 14(2): 143–159.

Finally, all the errors in this book are my sole responsibility and none of the sponsors or my co-authors on the above articles or my research collaborators on the research projects listed above are responsible for any of these errors.

Abbreviations

AF	Adaptation Fund
AHP	analytical hierarchy process
AOSIS	Alliance of Small Island States
BAU	business-as-usual
BRICS	Brazil, India, China and South Africa
CAP	Capital Asset Pricing
CDM	Clean Development Mechanism
CER	certified emissions reduction
CO_2	carbon dioxide
CO_2-E	carbon dioxide equivalent
DNA	Designated National Authorities
DPM	Deliberative, Pluralistic, Multi-Scalar theory of valuation
DTV	Dewey's Theory of Valuation
DU	Discounted Utility model
EU	European Union
FDI	foreign direct investment
GDP	gross domestic product
GEF	Global Environmental Facility
GHG	greenhouse gas
IAD	Institutional Analysis and Development framework
IEA	International Energy Agency
IMF	International Monetary Fund
IPCC	Intergovernmental Panel on Climate Change
IUCN	International Union for Conservation of Nature
JI	Joint Implementation
JUSCANZ	Japan, United States, Canada, Australia, and New Zealand alliance
KP	Kyoto Protocol
LUCF	land-use change and forestry
MADM	multiple attribute decision-making
MCDM	multiple criteria decision-making
MAUP	Modifiable Area Unit Problem
MODM	multiple objective decision-making
ODA	official development assistance

OPEC	Organization of the Petroleum Exporting Countries
REDD+	Reduced Emissions from Deforestation and Forest Degradation
SAW	simple additive weighting
SD	sustainable development
TEEB	The Economics of Ecosystems and Biodiversity
TEV	Total Economic Value
TOE	tons of oil equivalent
UNCTAD	United Nations Conference on Trade and Development
UNEP	United Nations Environment Programme
UNFCCC	United Nations Framework Convention on Climate Change
USNFIP	U.S. National Flood Insurance Program
VMT	vehicle mile(s) traveled
WTO	World Trade Organization

1 Introduction
Post-Kyoto climate governance

The lack of international action on human induced global climate change

The human induced global climate change is a complex challenge that will determine the future of human civilization and its place in the broader evolution of life on the planet earth. Despite persistent warnings from climate change scientists (IPCC 1995, 2001, 2007) and mounting evidence of disruptive climatic changes all over the planet (IPCC 2012), the international community of nations has failed to take meaningful action in proactively reducing greenhouse gas (GHG) emissions that are primarily responsible for human induced climate change. Based upon the agreements reached at the first Rio summit in 1992, the United Nations Framework Convention on Climate Change (UNFCCC) was given a rather vague mandate to prevent global climate change from reaching any "dangerous" levels. The lack of any democratic system at the international scale in this so-called "age of anthropocene" has further complicated the mandate of the UNFCCC.

The modest success attained by the UNFCCC process through the Kyoto Protocol (KP) in mandating approximately 5 percent reduction in the GHG emissions from the industrialized countries by 2012 below 1990 levels has been blown up due to the non-participation of the largest GHG emitting country (the United States) in ratifying the Kyoto Protocol, the withdrawl of another large GHG emitting country from the KP (Canada) after it realized it cannot meet its agreed upon GHG emission reductions and the issuance of so-called "hot air" emission credits to Russia and other eastern European countries. The lack of any meaningful mandatory GHG reductions from the so-called industrializing nations, such as the BRICS countries (Brazil, India, China, and South Africa) further weakened the potential of KP in reducing planetary scale GHG emissions. Consequently, human induced global climate change continues to take place largely unmitigated except for brief periods of global recession.

Amidst calls about the failure of Kyoto Protocol to regulate the growth of GHG emissions (Cass 2006; Harrison and Sundstrom 2007), a plethora of policy architectures have been proposed for designing an international post-Kyoto climate change governance regime. Bodansky *et al.* (2004), for example, provide an overview of sixty-four policy architectures that have been proposed in the recent literature. Different subsets of these sixty-four post-Kyoto policy architectures have been evaluated on the criteria of equity and justice (Baer 2002; Klinsky and Dowlatabadi 2009); fairness principles and operational requirements (Torvanger and Ringius 2002); environmental outcome, dynamic efficiency, cost effectiveness, equity, flexibility in the presence of new information, and incentives for participation and compliance (Aldy *et al.* 2003); cost-effectiveness, and compliance and participation (Barrett and Stavins 2003); regional representation of proposal authors (Kameyama 2004); quantitative and non-quantitative approaches (Philibert 2005); credibility, stability, flexibility, and inclusiveness (Biermann 2005); institutional design, participation, flexibility, and quantity versus prices (Guesnerie 2006); and technological cooperation (De Coninck *et al.* 2008). All of these criteria are perhaps critical in comparing post-Kyoto global climate governance and policy architectures, yet there are no constitutional, legal, institutional or moral reasons to prefer one set of evaluation criteria over the others. In other words, the complexity of orchestrating a global governance and policy regime poses fundamental challenges, as it has been demonstrated in the endless negotiation processes that have taken place without any meaningful action since the early 1990s.

The design of a post-Kyoto international climate governance regime under different policy architectures will, it has been argued and counter-argued during these negotiations, require a choice between binding versus non-binding and grandfathering versus per capita or some other complex decision heuristic. The Kyoto Treaty employed a grandfathering decision heuristic to set up binding emission reduction targets. A grandfathering decision heuristic establishes a baseline year (1990 for Kyoto) and participating nations agree on binding targets and deadlines to reduce GHG emissions to a certain percentage below the baseline year, which is a 5.2 percent below 1990 levels for certain developed, so-called Annex I, countries by 2008–2012 under the Kyoto Protocol (UNFCCC 1998: Appendix B). The grandfathering decision heuristic has been criticized by the overwhelming majority of developing nations (including large GHG polluters such as China and India) for granting a "free ride" to the industrialized nations for the GHG emissions that were either emitted prior to the baseline year since the onset of the Industrial Revolution in the 1750s or that are being emitted well above global per capita averages since the baseline year. Some of these GHG emissions will stay in the atmosphere for hundreds and thousands of years and, representatives of developing countries argued, any "equitable" GHG emission reduction plan must take into account these GHG emissions that were an unintended consequence of the industrialization process that brought overwhelming development to the industrialized world.

The "grandfathering" decision heuristic is not an acceptable option for many developing countries for a post-Kyoto climate governance regime. Instead of the grandfathering decision heuristic, GHG emission caps on a per capita basis appear more favorable to many developing country representatives of international climate policy negotiations because a per capita decision heuristic will arguably preserve their right of development. In contrast, many developed countries, especially the United States, Canada and Australia, have consistently shown their disagreement with the per capita decision heuristic. These developed countries argued that per capita decision heuristic will not only result in more stringent caps on the developed nations with higher historical GHG emissions, but it will also allow for larger populations of the developing countries. In fact, some of these high emitting countries, defined in this book as the countries that emit statistically significantly higher GHG/capita emissions than the global average for a predefined period, have even threatened to boycott any treaty that imposes binding GHG emission reductions in post-Kyoto phase. Were these high emitting countries to agree on a binding regime, the methodological dilemma surrounding the choice of one or another decision heuristic will require a resolution for designing a post-Kyoto governance regime.

An interdisciplinary and meta-theoretical perspective

The tension between developing and developed countries in the design of binding versus non-binding policy architectures and grandfathering versus per capita decision heuristics could be analyzed from a variety of theoretical and policy analytical perspectives. In the broader literature, these theoretical perspectives cut across environmental policy, ecological economics, climate science, environmental law, environmental psychology, political geography, political science, political ecology, political anthropology, international relations, and environmental sociology, which is by no means a comprehensive list. In this book, a multi-disciplinary and a meta-theoretical approach is deliberately adopted to present a multi-dimensional yet integrative critique of climate policy with a focus on meaningful policy and governance actions that must be taken in designing a post-Kyoto global climate governance and policy regime. From a meta-theoretical perspective, three theoretical perspectives are applied in this book to contrast and compare various elements of a post-Kyoto global climate governance regime: I characterize the first theoretical perspective as so-called "rational" perspective, primarily driven by game theoretical developments in computational sciences and international relations. Elinor Ostrom's (2005) Institutional Analysis and Development (IAD) framework is a slight modification of this framework, from a rational to a boundedly rational direction. In this theoretical perspective, actors at multiple levels are represented as players with their strategic, tactical, and operational decision-making choices. Strictly speaking, rational players are ruthless and cold, who calculate their expected utilities for alternate decision-making choices under different information regimes. Since global climate change mitigation and adaptation policy issues are beset with risk and uncertainty, interactive decision-making by rational (and boundedly rational) actors could be modeled/explained under conditions of risk and uncertainty. This theoretical perspective seeks to elicit specific conditions under which international scale cooperation among different competitive players could/should emerge. Further, appropriate institutional designs that could/should emerge to solve environmental externality problems are explicitly derived through this theoretical framework.

I characterize a second theoretical perspective as "constructivist" and "social psychological." Under this second perspective, game theoretical descriptions and/or prescriptions are assessed through critical policy analytical lenses that are used to analyze the complex challenge of designing an international post-Kyoto climate policy regime. In particular, these critical policy analytical lenses are labeled as the politics of scale; the politics of ideology; and the politics of knowledge. In the context of international climate policy, the politics of knowledge lens, for example, analyzes what type of knowledge "hegemonizes" the discourse for designing climate change mitigation and adaptation policies. In particular, politics of knowledge deeply affect the foundational policy and governance design questions concerning government, market, and society relationships. Marketization of climate governance can result in undermining governmental and societal goals through politics of knowledge. Further, knowledge "hegemony" can also influence the social construction of accountability and adaptation in global climate governance. Differential considerations of the knowledge could derive different conclusions about the design of policy mechanisms, binding versus non-binding emission reduction commitments and specific decision heuristics to allocate the emission entitlements. The politics of knowledge lens sheds light on such climate policy disputes and elucidates the politicized nature of climate science. In particular, the role of policy narratives and knowledge hegemony during international negotiations could be assessed through politics of knowledge policy analytical lenses.

Two connected aspects of hegemony in critical policy studies are emphasized:

> On the one hand, hegemony is a kind of political practice that captures the making and breaking of political projects and discourse coalitions. But on the other hand it is also a form of rule or governance that speaks to the maintenance of the policies, practices and regimes that are formed by such forces.
>
> (Howarth 2009: 301)

Both of these connected aspects of hegemony are explored in this book in the context of UNFCCC driven climate treaty negotiation process and politics of knowledge to set up and design a global climate governance regime. It is found in the negotiation data that hegemonic practices are utilized by high emitting countries through politics of knowledge in such a way that their proposed policy regimes are used to *conceal* the historical and future responsibility of high emitters and to *naturalize* their proposed policy architectures and decision heuristics. Using a critical policy analytical lens of politics of knowledge, this second theoretical perspective enables me to explain why high GHG emitting countries generally prefer non-binding emission reduction targets and grandfathering during the UNFCCC negotiations. Further, it enabled me to explore whether grandfathering and other types of non-binding post-Kyoto policy architectures are being *naturalized* in the international climate policy discourse to *conceal* the historical and future responsibility of high GHG emitters.

The notion of binding commitments has been strongly contested among various coalitions of countries in the UNFCCC negotiation process. Figure 1.1, as a demonstrative example, shows five negotiating blocks and their respective positions vis-à-vis binding versus non-binding GHG emission reduction commitments from a qualitative assessment of UNFCCC negotiation data from 1995 through 2010. The G-77 and China consistently negotiated for binding GHG commitments for industrialized countries. The group of Organization of the Petroleum Exporting Countries (OPEC), on the one hand, consistently opposed any binding GHG emission reduction commitments, while the group of Alliance of Small Island States (AOSIS), on the other hand, consistently supported binding GHG emission reduction commitments for all countries. European Union (EU)-15 countries and later EU-25 countries changed their positions from binding commitments for only industrialized countries to all the countries. The position of Japan, United States, Canada, Australia, and New Zealand (JUSCANZ) vacillated over the years from no binding commitments for any country to binding commitments for all countries to just the binding commitments for industrialized countries, as shown in Figure 1.1.

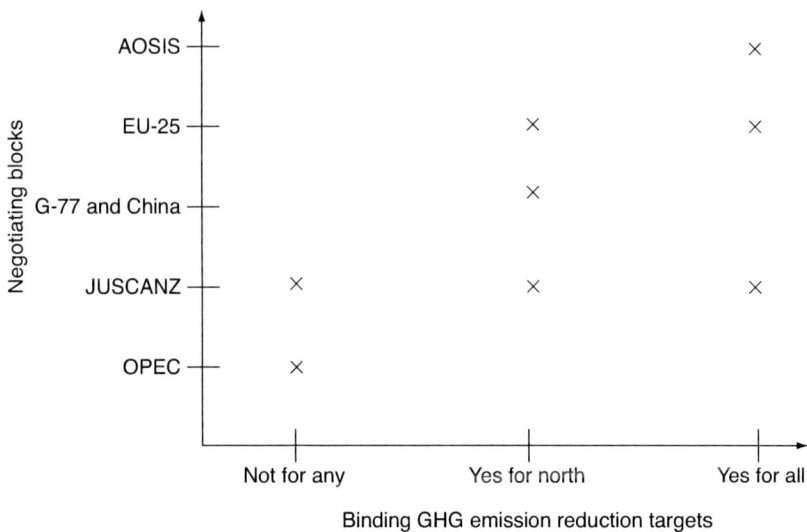

Figure 1.1 The historical position of negotiating blocs in UNFCCC process vis-à-vis binding GHG emission reduction targets.

Notes
AOSIS: Alliance of Small Island States; EU-25: European Union; G-77 and China: group of about 139 developing countries; JUSCANZ: Japan, United States, Canada, Australia, and New Zealand; OPEC: members of the Organization of Petroleum Exporting Countries.

From rational game theoretical and boundedly rational institutional design perspectives (Luterbacher and Sprinz 2001; Cass 2006), it can be hypothesized that the coalitions that stand to lose most from binding commitments (e.g., OPEC and JUSCANZ) are opposed to any notion of binding commitments for emission reductions. On the other hand, the coalitions that stand to lose most from inaction (i.e., non-binding commitments), such as AOSIS, are in strong support of policy mechanisms that require binding commitments for a post-Kyoto Treaty.

In addition to binding versus non-binding emission reduction commitments, another important question in post-Kyoto climate policy discourse, especially multi level governance, concerns whether some universal decision heuristic must be necessarily used for a top-down allocation of emission entitlements or a bottom-up open-ended process be allowed to govern the emergence of binding or non-binding commitments at sub- or supra levels of nation states. Figure 1.2 shows twenty-two out of the sixty-four proposed post-Kyoto climate policy architectures categorized on the criteria of binding/non-binding emission reductions and top-down versus bottom-up policy development designs. Authors who primarily proposed these twenty-two policy architectures are mentioned in numerical order and relevant citations are shown at the bottom of Figure 1.2.

From a "constructivist" and social psychological perspective, It could be argued that the policy architectures that promote non-binding emission allowances, whether top-down or bottom-up, *conceal* avoidance of climate change problematic by *naturalizing* the discourse that nation-states may never be able to arrive at an effective and equitable international climate treaty. Some of these bottom-up and non-binding policy architectures shown in the south-eastern quadrant of Figure 1.2, such as sustainable development policies and measures, portfolio approach, human development goals, converging markets, and climate marshal plan, are probably noble in spirit, but game theoreticians would readily identify that these policy architectures are vague, at best, and may result in climate change mitigation outcomes that are worse than business-as-usual scenario (e.g., 750 to 1000 parts per million (ppm) carbon dioxide (CO_2) concentrations by the end of the twenty-first century). While implementation of sustainable development policies and climate Marshall Plans might enable global climate change mitigation and adaptation, the non-binding nature of underlying emission commitments could also translate into non-binding financial and technological commitments for sustainable development policies and Marshall Plans. Allocation of GHG allowances in non-binding policy architectures is a non-starter, and, in fact, can be demonstrated to be a winning strategy for high emitting negotiating blocs such as OPEC and JUSCANZ.

Similarly, top-down and non-binding policy architectures in the north-eastern quadrant of Figure 1.2, such as agreed domestic carbon taxes, international agreements on energy efficiency and north-south dialogue, entail climate policy prescriptions that might enable GHG polluting countries to agree on some internationally negotiated adjustments in energy, agriculture, and other GHG related sectors without requiring these countries to actually implement these multi-sector adjustments. Policy architectures in the category of binding commitments agreed through a bottom up process, as shown in the south-westerly quadrant of Figure 1.2, provide interesting test bed for the so-called "neo-liberal institutionists" who believe in the charitable leadership of high GHG polluters (e.g., EU) in committing to binding restrictions. The neo-liberals, however, do not seem to question the level or basis of binding commitments. The proposal of bottom-up national commitments, for example, will leave it up to the national governments to come up with some random number, depending upon their specific context and climate

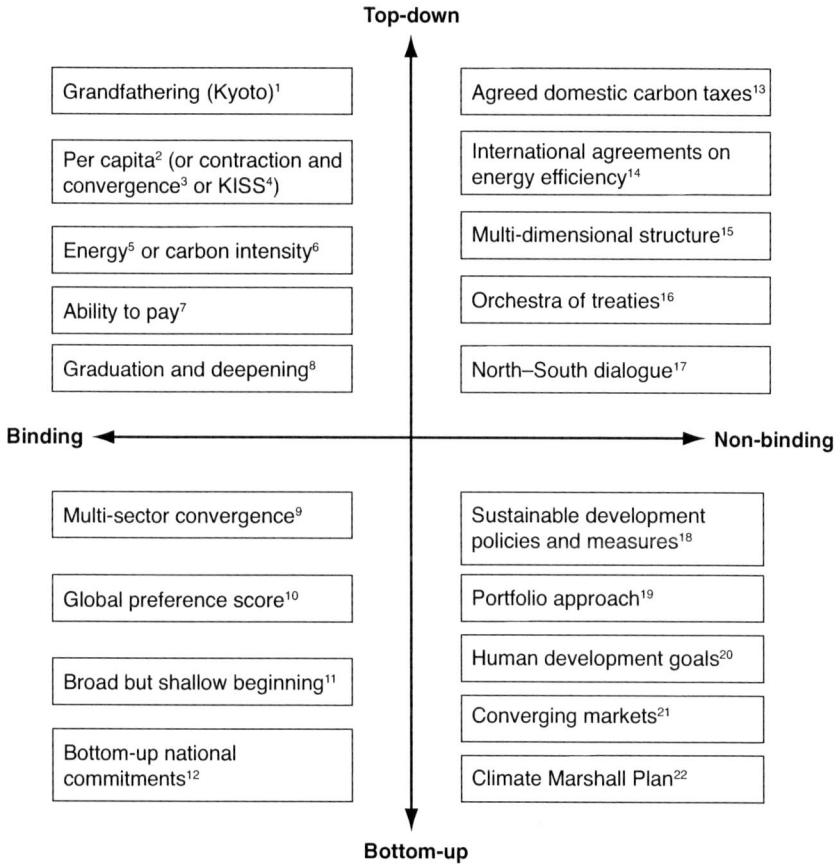

Top-down

Binding		Non-binding
Grandfathering (Kyoto)[1]	Agreed domestic carbon taxes[13]	
Per capita[2] (or contraction and convergence[3] or KISS[4])	International agreements on energy efficiency[14]	
Energy[5] or carbon intensity[6]	Multi-dimensional structure[15]	
Ability to pay[7]	Orchestra of treaties[16]	
Graduation and deepening[8]	North–South dialogue[17]	
Multi-sector convergence[9]	Sustainable development policies and measures[18]	
Global preference score[10]	Portfolio approach[19]	
Broad but shallow beginning[11]	Human development goals[20]	
	Converging markets[21]	
Bottom-up national commitments[12]	Climate Marshall Plan[22]	

Bottom-up

Figure 1.2 Categorization of post-Kyoto policy architectures along binding and non-binding, and top-down and bottom-up dimensions (sources: 1. (UNFCCC 1998); 2. (Agarwal n.d.; Bode 2004); 3. (Meyer); 4. (Gupta 2003); 5. (Schmidt *et al.* 2008); 6. (Kim and Baumert 2002); 7. (Jacoby *et al.* 1998, 1999); 8. (Michaelowa *et al.* 2005); 9. (Sijm *et al.* 2001); 10. (Muller 1999, 2001); 11. (Schmalensee 1996); 12. (Reinstein 2004); 13. (Cooper 1998,2001); 14. (Ninomiya 2003); 15. (METI 2003); 16. (Sugiyama *et al.* 2003); 17. (Ott et al. 2004); 18. (Winkler *et al.* 2002); 19. (Benedick 2001); 20. (Pan 2003); 21. (Tangen and Hasselknippe 2005); 22. (Schelling 1997, 2002)).

change leadership, to agree to commit to GHG emission reductions. The policy architecture for broad but shallow beginning is a bit different: it asks countries to commit to reduce a very small fraction of what is needed, say for a 450 parts per million CO_2 constrained world, so the emphasis is on broader participation without requiring deep adjustments with the neo-liberal hope that relevant markets will get the signal and enable adjustment processes in the long run.

Overall, a meta-theoretical evaluation of the policy architectures in non-binding and bottom-up; non-binding and top-down; and binding and bottom-up categories may lead an evaluator to conclude that the policy architectures proposed in these three categories may provide substantial leverage in accommodating climate change mitigation and adaptation; however, it is equally likely that these architectures may not provide substantial leverage in weaning off the business-as-usual path to a much lower GHG emission path that might be needed for a 450 ppm or even 350 ppm CO_2 constrained world. Due to this uncertainty in commitment levels and potential lack of follow-through in the policy architectures proposed in these three quadrants of Figure 1.2, it is argued in this book that the policy architectures that are in the category of binding commitments with some kind of top-down arrangement, i.e., north-eastern quadrant in Figure 1.2, could provide more "effective, verifiable, measurable and quantifiable" international climate policy goals. Later chapters of the book explore this line of argumentation in more detail with the analysis of GHG emission data.

In particular, detailed evaluation of the policy architectures that are in the binding and top-down category is undertaken with respect to their implications on medium to long-term GHG emission entitlements across the nations. Defenders of grandfathering, for example, in the past have argued that it allows high GHG emitters to take into account the historical responsibility, so developing countries are not burdened with binding GHG emission reductions, as in Kyoto. In response, the opponents of grandfathering typically take a futuristic stance: since developing countries are expected to contribute a major portion of GHGs in the next 50 years, it is argued, grandfathering, by not requiring developing countries to agree on binding commitments, places a higher burden on the developed world by requiring it to bear the upfront burden of emission reductions. The implications of such debate in terms of GHG entitlements are explored through the evidence at hand.

Similarly, the policy architecture of graduation and deepening, which is markedly different from grandfathering policy architecture, entails that developing countries that cross pre-decided income thresholds will be eligible to be included (i.e., graduated) for deeper emission cuts. Nevertheless, the decision about income threshold and the "depth" level of emission reductions will probably require extensive negotiations to assess GHG emission entitlements. The ability to pay, yet another decision rule included in Figure 1.2, will also require similar income threshold criteria for assessment of emission allowances.

Energy or carbon intensity targets also present differentiated targets as developing countries, due to their early development stage, tend to have higher energy or carbon intensities, while developed countries who are responsible for the bulk of GHG emissions so far tend to have lower energy or carbon intensities due to their technological superiority. Setting up energy or carbon intensity targets will be like incentivizing the developed countries to continue on business-as-usual while requiring developing countries to leapfrog into a future that has costlier low energy and carbon intensities. The choice of one or another decision heuristic, no matter which one, leads to differential emergence of winners and losers in terms of GHG emission entitlements. The powerful industrialized countries have naturalized the usage of grandfathering as a decision heuristic in UNFCCC negotiations, the constructivists could argue, precisely because it offers them an advantageous positions in terms of avoiding larger emission reduction commitments under some of the other policy architectures shown in the north-western quadrant of Figure 1.2.

Third, I characterize complex systems' theoretical perspective as a so-called "complexity" perspective to assess and design various elements of an effective post-Kyoto global climate governance regime. From the complexity perspective, it has been argued that the "scientific management" view of governing social ecological systems, of which atmosphere is but one component, and consequent arrangement of international organizations could be misguided, and even dangerous, to adequately cope with global climate change, food insecurity, biodiversity loss and other such public policy crises. The complexity theorists question one of the central assumptions in top-down scientific management models of governing social ecological systems. This assumption states that the strategy spaces of players in dynamically evolving social ecological systems could be predefined to design institutions and their organizational functions. The recent evolution of international climate policy mechanisms, such as Reduced Emissions from Deforestation and Forest Degradation (REDD+), and biodiversity conservation policy mechanisms, such as The Economics of Ecosystems and Biodiversity (TEEB), provides examples of how UNFCCC and United Nations Environment Programme (UNEP) design "command and control" type of public policies that require predefinition of the strategy spaces of all players in the dynamically evolving social ecological systems. Later in the book, I draw meta-theoretical implications from a synthesis of recent REDD+ and TEEB mechanism design discourses to argue for a shift in policy design approaches from "Pareto Optimization" and predefined "Fitness Landscapes" to a complex systems perspective. A complex systems approach to the governance of social ecological systems could provide alternative ways to design the policies that govern global atmosphere, planetary oceans and forests. Under this "complexity" view, human civilization is called upon to be open to the partially unknowable Adjacent Possible, not to control, but to enable and adapt, and partially shape what will emerge. Furthermore, the social ecological systems could transition in and out of multiple stable states, or even exist far from equilibrium. A complex systems perspective could open up the discussion for many alternate policy designs, including proactive management of critical phase transitions in social ecological systems at local scales, decentralized policy-making, self-organization from local to global scales, and adaptive, democratically-anchored, international policy-making.

Since the Enlightenment, human civilization has increasingly lived with the presumption that via reason and knowledge, we would have the *power to manage*. It is the so-called "General Motors" view of scientific management from 1950 onwards. In this book, I am arguing that it is a profoundly inadequate model for how businesses, societies, civilizations and social ecological systems evolve. If one looks at the United Nations and other post-World War II Bretton-Woods Institutions such as the World Bank and International Monetary Fund (IMF), there seems to have been, since its inception, the same mid twentieth century faith that such management might even occur with some eventual form of a centralized world governance structure that would "know" and "allocate" and "manage." I am arguing in this book that this view of governing social ecological systems and the consequent arrangement of international organizations is misguided, and even dangerous, to adequately cope with global climate change, food insecurity, biodiversity loss and other emergent crises that pertain to the governance of social ecological systems.

My argumentation is based on questioning one of the central assumptions in top-down scientific management models of social ecological systems' governance. This assumption states that the phase space of dynamically evolving social ecological systems could be predefined to design institutions and their organizational functions. In contrast, I am arguing that we cannot pre-state what new "salients" of economic, ecological, social or civilizational activities will arise, from what little or large triggers. Further, we cannot optimize over a predefined phase space as a well-formulated decision problem. Instead, I argue that a complex systems approach to the governance of social ecological systems might provide an alternative way to design the functions and capacities of international organizations. Under this view, we are called upon to be open to the partially unknowable Adjacent Possible, not to control, but to enable and adapt, and partially shape what will emerge. This is the profound opposite from General Motors 1950, from dictatorship, from a controlling governance managed by the elites who "know." Under this complex systems perspective, we begin to see democracy in a new light, a framework of freedoms and procedures that allows and enables adaptability and shaping of the possibilities where we will go. I elaborate this perspective in the light of specific triple global crises currently known as global climate change, global food insecurity and global biodiversity loss, all of which are somehow linked with the anthropogenic intervention in social ecological systems.

Emphasis on maximizing economic efficiency for global production of goods and services has, for example, led to strongly centralized "free trade policies" that are enforced by international organizations such as WTO. Foreign direct investments (FDI) are promoted under such free trade regimes. Environmental impacts of such free trade policies and FDIs have not been internalized through imposition of Pigovian taxes in the conduct of international trade. The powerful industrial North and transnational corporations have been able to negotiate terms of trade that maximize their profits and retain current socio-political power patterns, while greenhouse gas (GHG) emissions resulting from global industrial complex, mining, deforestation, wasteful consumption, and fossil energy burning continue to accumulate in coupled atmospheric and oceanic systems. Global climate change science is unable to predict an accurate timing of critical phase transition; however, all global circulation models agree that the business-as-usual path of GHG emission trajectory will sooner or later cause a phase transition in the coupled atmospheric system, after which socio-political policy actions and behavioral changes by themselves will not be adequate to stop run-away climate change because the natural biogeochemical cycle would have degenerated to the point that reduction in anthropogenic GHG emissions would be inadequate to stop the global warming effect from playing havoc in diverse socio-ecological systems. Protection of free trade through centralized global institutions and maximization of economic growth will only exacerbate GHG emissions. On the other hand, international climate policy experts are well aware of the fact that any strong action by a handful of nation-states would shift global industrial production to the nation-states that do not take any meaningful action on regulating GHG emissions. In technical jargon, this issue is known as the "leakage" problem.

In this book, I thus argue that the management of global climate, food and biodiversity crises pose a fundamental *normative* or *value ambiguity* challenge, i.e., experts in centralized international organizations do *not* and can *not* know the phase space of variables and strategies over which the optimization decision problem is to be stated. To elaborate this further, I discuss the global crisis of human-induced global climate change as a value ambiguity problem. Both the mitigation and adaptation to global climate change requires deep, long-term foresight, and unwavering collective/normative human action at the global scale. Yet, the complexity of managing global climatic change, and its impacts on irreversibly changing the evolutionary pathways of biological, technological and economic systems remains a deep puzzle, beyond the reach of positive sciences. From a normative standpoint, which is an essential component of any governance effort to deal with the complexity of complex systems, I argue in this book that the international organizations need to move beyond the positivistic goals of managing and controlling global social ecological systems, and that management and international organization sciences need to move beyond "optimization-envy." I argue that there is a need to accommodate both facts (understanding) and values (normative prescriptions) for managing global environmental and social crises. Furthermore, international organizations (e.g., UNFCCC, UNEP, WTO) will need to let go of reductionism, and replace it with the acknowledgment of normative (or value-laden) complexity of governing complexity, which in turn will require a shift from expert-based international organizations to democratically-anchored governance networks, a core concept in terms of institutionalizing accountability in the global climate governance regimes.

From the synthesis of REDD+, TEEB and WTO policy designs, it is obvious that: first, there are many stakeholders; second, different stakeholders have differential power; and, third, different stakeholders have different "valuation" or utility functions that are often in conflict, with the powerful winning. Reframing these conclusions formally, let us consider a continuous policy space. Now let us take any one stakeholder and its utility function over the policy space. This yields a "fitness landscape" or "payoff landscape." Let us do the same for each different stakeholder and we have N landscapes over the strategy space for N stakeholders. In general these N fitness landscapes will *not* have global optima, or even local peaks, at the same locations on the strategy space. Now, as I pointed out above, these diverse landscapes, plus diverse power capacities from the World Bank, to local hunter-gathering peoples, yield a complex "tug of war" about policy. A natural solution concept in the above N landscape case, where *we cannot say the relative values of the N different utility functions of the N stakeholders*, is that of global Pareto solutions. A global Pareto solution is a point on the landscape such that no local move in a continuous phase space can increase utility for one stakeholder or individual, without decreasing at least the utility for somebody else in the bounded system.

Formally, given the set of N stakeholders, a game in normal form could specify a strategy set (i.e., phase space) for each stakeholder. For stakeholder j, we denote by S^j the strategy space for the jth stakeholder; and define $S := \Pi_{j \in N} S^j$ as the strategy space for all N stakeholders. Further, it is postulated that a preference relation among the strategy spaces for jth stakeholder is represented by a utility function $u^j : S \times S^j \to \mathbf{R}$. This implies that when all the stakeholders have chosen their strategies $s := (s^1, \ldots, s^n) \in S$, stakeholder j will enjoy the utility level $u^j(s, \xi^j)$ when a stakeholder changes a strategy from x^j to $\xi^j \in S$. A game among N stakeholders in normal form could now be defined as a list of specified data $\{S^j, u^j\}_{j \in N}$. A Nash equilibrium of a such a stakeholder game in normal form $\{S^j, u^j\}_{j \in N}$ is an n-tuple of strategies $s^* := (s^{1^*}, \ldots, s^{n^*}) \in S$ such that for every $j \in N$, $u^j(s^*, s^{j^*}) \geq u^j(s^*, \xi^j)$ for all $\xi^j \in S$. In standard social welfare analysis, the real number $u^j(s^*, s^{j^*})$ is the utility that jth stakeholder currently enjoys, and in equilibrium there is no incentive for her to change her strategy s^{j^*} by herself. In general, Nash equilibrium of such a game is considered as a *descriptive* concept, i.e., it describes what the "fitness landscape" would be when N stakeholders play the game, assuming that each stakeholder has a nonempty, convex, compact subset of strategy space, and a continuous quasi-concave utility function. In contrast to a descriptive concept, a *normative* concept such as Pareto optimality is typically defined to signify a desirable "fitness landscape" or outcome space of a game. These normative concepts, such as Pareto optimality, are typically used as "targets" by the "General Motor" style rational planners situated in the World Bank, WTO and other international organizations. This in turn leads to the emphasis on "win-win-win" rhetoric in the creation of international policies such as REDD+ and TEEB described above. Formally, a Pareto optimal strategy bundle $s^* \in S$ for a game in normal form $\{S^j, u^j\}_{j \in N}$ could be defined by stipulating the condition that if it is not true that there exists $v \in S$ such that $u^j(v, v^j) \geq u^j(s^*, s^{j^*})$ for every $j \in N$. It can be demonstrated in environmental governance conundrums such as REDD+ and TEEB that the Nash equilibrium of prisoner dilemma type of games is not necessarily Pareto optimal. Institutional political economists that support current REDD+ and TEEB policy mechanisms would thus argue that these mechanisms would move the global governance of tropical forests from less optimal to more Pareto optimal situations; and by implication, fitter landscapes, in the long run, by incentivizing the developing countries to conserve tropical forests, protect biodiversity and maintain carbon stocks.

Given the policy design challenges about REDD+ and TEEB described above, as well as generic crisis facing international organizations in terms of dealing with global climate change, global food insecurity as well as global bio-diversity loss, I argue in this book that such rationalist and normative arguments based upon the logic of Pareto optimality and Nash equilibria lead to poorly defined international institutions that create perverse incentives for local and indigenous communities, as defined for REDD+ and TEEB above, displace bio-diversity through the removal of old growth forests, engender inequities due to century old property right and tenure conflicts, assume technological methodolo-gies that cannot objectively assign baselines, and above all, place a monetary value on natural and biological systems that trivializes the worth of biodiversity and social-ecological systems through assumption-laden game theoretical and institutional design frameworks.

Due to these invalid assumptions, and lack of reality-checks, I argue that Pareto optimal fitness landscapes could not be predefined by rational and/or cen-tralized planners. This in turn implies that the phase spaces of dynamic social ecological systems could not be pre-defined by rational planners. In fact, from a long-term inter-temporal perspective, I argue that the strategy spaces and utility functions of different stakeholder groups over N policy spaces could not be pre-stated. I emphasize this unprestatibility by arguing that the utility functions and strategy spaces of different stakeholder groups are not *fixed*; rather they are very highly context dependent. The utility functions and strategy spaces of different stakeholders change with changes in technology, boundary conditions, biologi-cal evolution and other endogenous and exogenous drivers of change in the social-ecological systems that are typically ignored when modeled by Pareto optimizing rational planners (e.g., Stern 2008). In the context of REDD+ and TEEB policy designs, if we consider a scenario where the global food prices are tripled, this will indirectly also triple the opportunity costs for REDD+ and TEEB payments for developed countries, which might change optimal decision in favor of other carbon abatement technologies. In another scenario, a techno-logical innovation (e.g., carbon sequestration from coal fired power plants and/or more cost effective production of solar cells) could change the optimal strategies for GHG polluting countries that could dry up REDD+ payments. Further, in a more cynical scenario, global climate change could initiate perverse positive feedback loops through unintended consequences, whereby higher CO_2 in the atmosphere could lead to less precipitation in the tropical forested systems, which in turn could lead to precipitous decline in forest systems, which in turn could speed up the concentration of CO_2 in coupled atmospheric and oceanic systems, further aggravating the global carbon cycle. The capacity of tropical forests to absorb CO_2 under high CO_2 concentration scenarios is still a very contested and uncertain issue among biology and ecology experts. Given these economic, technological and biological uncertainties in future scenarios, predefining inter-temporal phase spaces of tropical social ecological systems and acting upon them through international institutional mechanisms such as REDD+ and TEEB appears to contain many uncertainties full of unintended consequences.

Instead, I argue that international organizations need to incorporate a complex systems perspective in designing and supporting international policy and institutional mechanisms. Under a complex systems perspective, social-ecologoical systems could transition in and out of multiple stable states (Scheffer 2009), or even exist far from equilibrium (Kauffman 1993, 1995). Instead of arguing over the expected opportunity costs, a complex systems perspective, which focuses upon "adjacent possible" while keeping larger scale impacts of policies, institutions and international organizations in the context, could open up the discussion of many alternate policy designs, including proactive management of critical phase transitions in social-ecological systems at local scales, decentralized policy-making, self-organization from local to global scales, and adaptive policy-making.

An adaptive, decentralized and democratically anchored global governance of tropical forests could, for example, provide adequate voice and empower the indigenous communities, and restrain national governments and international organizations from trading off old-growth tropical forests and biodiversity for the sake of maintaining global carbon cycle. Rich industrialized countries could rather focus on their local carbon footprints and not use elusive carbon offsets from tropical forests to sustain their ecologically disrespectful lifestyles. Furthermore, international organizations, such as WTO, need to be fundamentally reformed so that they do not incentivize destruction of tropical forests through the ideological knowledge of marketization of tropical forests and free trade of minerals and soils available in tropical forests. In fact, inter-organizational coordination among international organizations could prevent such conflicting policies, where WTO is promoting deforestation, while UN-REDD and UNEP programs are promoting forest conservation; where the World Bank is promoting unfettered economic development, while the International Union for Conservation of Nature (IUCN) is promoting biodiversity conservation; where IMF is promoting globalization and corporatization, while global capacity to sustain such globalization is diminishing faster than anticipated by many international development experts.

Transitioning from Kyoto to post-Kyoto climate governance

There are two starkly different perspectives on the legacy left behind by Kyoto Protocol for a post-2012 climate governance regime. On the one hand, we have a group of experts who see the glass half full. They argue that global climate change is a global scale problem that has no historical precedent. The regulation of ozone depleting gasses through Montreal Protocol appears closest in form, yet the complexity of dealing with climate change is much higher and unprecedented. From the perspective of these optimistic experts, first steps in setting up global governance and policy regimes are necessarily incremental. The Kyoto Protocol thus laid the foundations for a global climate policy regime that will be further strengthened in a post-Kyoto timeframe, taking a medium- to long-term view as commensurate with the long-term climate problem. Second, on the other hand, we have experts who frame the legacy of Kyoto Protocol as essentially a failed experiment. For these experts, the Kyoto experiment demonstrated the failure of working within the capitalistic system that caused the climate problem in the first place. These experts criticize marketization of climate governance and, for supporting their claim, argue that Clean Development Mechanism (CDM), Joint Implementation (JI) and REDD have had barely any meaningful effects on energy or forestry sectors in reducing GHG emissions.

As we transition from Kyoto to a post-Kyoto governance regime, the meta-theoretical perspective taken in this book necessitates acknowledgement of both of these perspectives. The rationalists, game theoreticians and policy regime theorists appear optimistic about the legacy of KP from a medium- to long-term perspective. On the other hand, the constructivists, social psychologists and critical theorists/political ecologists do not seem to consider KP's legacy as very useful in terms of mitigating anthropogenic climate change. It is in this latter context, the context of the constructivists, social psychologists and critical theorists/political ecologists that I have broadly framed the argumentation in this book to inform the debate about transitioning from a KP to a post-Kyoto global climate governance regime. In particular, I use three critical policy analytical lenses to analyze various policy and governance architectures for a post-Kyoto world. I call these lenses as representing the politics of scale, the politics of ideology and the politics of knowledge. The politics of scale lens focuses on the theme of temporal and spatial discounting observed in human societies and how it impacts the allocation of environmental commons and natural resources across space and time. The politics of ideology lens focuses on the themes of risk and uncertainty perception in complex, pluralistic human societies. The politics of knowledge lens focuses on the themes of knowledge and power in terms of governance and policy designs.

It is argued throughout the book that the first perspective (optimist institutional designers, rationalists, game theoreticians, policy regime theorists) will fail to effectively deal with the climate change problem until and unless the politics of scale, politics of ideology and politics of knowledge are explicitly confronted in designing post-Kyoto climate change mitigation and adaptation policies and governance regimes. Indirectly, I'm thus arguing that constructivists, social psychologists and critical theorists/political ecologists also deserve a seat on the governance design table. Further, since global climate change is a complex problem, beset with uncertainties and sheer ignorance, complexity theorists might also inform the debate. Especially, those complexity theorists who are studying coupled natural and human system and/or social ecological systems can bring new insights in informing this global policy discourse. Overall, thus, this book presents a meta-theoretical perspective, the perspective of rationalists, constructivists and complexity theorists, to inform the ongoing discussions and negotiations surrounding the design of post-Kyoto governance and policy regimes.

The road map of this book

Broadly, Chapters 2 and 3 present analytical findings with a focus on the politics of scale; Chapters 4 and 5 focus on the politics of ideology; and Chapters 6 and 7 focus on the politics of knowledge in terms of designing a post-Kyoto climate governance regime. The politics of scale, ideology and knowledge are, however, explicated in a meta-theoretical context through the presentation of rationalists, constructivists and complexity theorists' positions about various climate governance and policy design related conflicting and consensual issues.

More specifically, in Chapter 2, an in-depth analysis of global climate change mitigation policy, especially the allocation of greenhouse gas emission allowances, is undertaken to elucidate the repercussions of various policy designs across spatial and temporal scales. Further, in Chapter 2, the politics of scale lens is deployed to analyze climate change adaptation policy. In particular, the problem of designating "risk zones" for adaptation policy instruments, such as risk insurance, and the distribution of risk across present and future generations is analyzed. Next, in Chapter 3, an adaptive management framework is developed to assess synergies and trade-offs across multiple space-time scales that are caused by policy interventions in complex systems. This framework is applied to re-contextualize the current debate surrounding the design and governance of climate change mitigation policy mechanisms in multiple space-time scales. One specific policy mechanism, Reduced Emissions from Deforestation and Forest Degradation (REDD+), which is being negotiated for a post-Kyoto climate treaty, is analyzed from the perspective of adaptive management framework to identify cross-scale synergies and trade-offs that are faced with respect to climate change mitigation vis-à-vis biodiversity conservation and food security. Further, CDM, another KP policy mechanism that is expected to be part of a post-Kyoto governance regime, is also evaluated through the lens of politics of scale.

Next, in-depth analysis is provided in Chapter 4 to explicate the impact of ideology on the public understanding of climate change science and the pervasive psychology of denial that is observed with respect to the implementation of potential climate policy mechanisms. In particular, the politics of ideology lens is applied to explain the psychology of denial that is observed in many climate change risk assessment studies. Further, in Chapter 4, ideological driven political gridlock, driven by a variety of politicians, corporations, mass media and citizens, is analyzed from the perspective of politics of ideology lens. Building upon the analysis in Chapter 4, in Chapter 5, the challenges of communicating uncertain climate risk in the face of politics of ideology are described in the larger context of behavioral psychological and experimental decision theoretical research. Specific proposals to reframe climate science communication that cut across ideological gridlock are also addressed in Chapter 5.

The politics of knowledge lens is deployed in Chapter 6 to analyze the policy debates that surround the foundational policy and governance design questions concerning government, market and society relations for setting up effective and fair climate mitigation and adaptation policies. In particular, marketization of climate governance that was attempted in the KP through the so-called flexibility mechanisms (CDM, JI and later on REDD+) is addressed from the perspective of politics of knowledge and implications are drawn for a post-Kyoto climate governance regime. Further, Chapter 7 continues to apply the politics of knowledge lens and elucidates the role of "hegemonic" knowledge in the social construction of accountability across governance networks. Challenges pertaining to inter-generational and intra-generational allocation of GHG emission entitlements are addressed in this context. Further, the emergent design of Adaptation Fund (AF) is also evaluated from the perspective of politics of knowledge lens in Chapter 7. Finally, in Chapter 8, the book concludes with the broader environmental and sustainable development issues that require governance of environmental, social and political complexity. Various proposals about confronting the politics of scale, ideology and knowledge with respect to post Kyoto Climate governance design are explored and their potential impacts on the emergent development pathways for future generations are briefly assessed.

2 The politics of scale I
Temporal and spatial discounting

Temporal discounting and inter-generational allocation problems

Since the World Commission on Environment and Development (WCED 1987) conceptualized sustainable development as development that meets the needs of the *present* without compromising the ability of the *future* generations to meet their own needs, almost all definitions of sustainability implicitly or explicitly address issues of inter-generational allocation of environmental resources (Norton 2005). Allocation of different kinds of natural capital over time, for example, signifies resource allocation decisions that will directly affect inter-generational environmental sustainability. While different notions of sustainability and sustainable development have been adopted in the key policy goals of different international agencies (e.g., UN, IUCN), the initial enchantment of national governments, and NGOs since WCED (1987), with sustainable development as a "win-win" panacea has given way to the emerging notion of "hard choices" and "difficult trade-offs" that entail inter-generational allocation of environmental resources (Ostrom 2007; McShane *et al.* 2011).

In this section, I focus on explaining inter-temporal value trade-offs that are inherent in sustainable management of natural systems that include global atmospheric and oceanic systems. In particular, I focus on a specific inter-disciplinary theoretical tension that exists between two camps of sustainability theorists in operationalizing inter-temporal value trade-offs. The first camp, predominantly represented by neoclassical economic theorists and their offshoots, argues for positive discount rates in comparing the costs and benefits of inter-temporal resource allocation decisions (for example see Becker 1976, 1993; Solow 1993; Beckerman 1994; Nordhaus 1994; Stern 2008). The second camp, predominantly represented by systems analysts, decision scientists and behavioral scientists, argue for an open-ended elicitation of inter-temporal value trade-offs, which implies that decision-makers could display negative, zero, positive or even non-linear discount rates on different sets of values in different decision contexts (for example, see Norton 1987, 1989, 1991, 1994, 1996, 2000, 2005; Toman 1994; Norton and Toman 1997; Norton *et al.* 1998; Gilovich *et al.* 2002; Keeney 2002; Loewenstein 2003; Jacobi and Hobbs 2007; Hämäläinen and Alaja 2008; Ariely 2009).

Neo-classical economic theory frames the assessment of inter-temporal value trade-offs from the normative perspective of a Discounted Utility (DU) model, which was initially postulated by Samuelson (1937). The DU model posits that people have a single unitary rate of time preference that they use to discount the value of delayed/future events. Toman (1994: 400) succinctly presents the dilemma for inter-generational equity and sustainability posed by positive discounting inherent in the DU model:

> The typical criterion of discounted inter-temporal welfare maximization in applied welfare economics occupies one point in the continuum of alternative justice conceptions. This criterion not only emphasizes preference satisfaction over rights; it also is highly presentist, since with any positive intergenerational discount rate the welfare of individuals living one generation in the future is scarcely relevant to current decision-making. Many writers have suggested that the presentist focus of the present-value (PV) criterion implies an influence of the current generation over the circumstances of its more distant descendants that seems, at least intuitively, to be ethically questionable.

Notwithstanding the presentist bias in the DU model, it is widely used in cost benefit analysis, total economic valuation, valuation of ecosystem services (for example see Freeman 2003), and intertemporal GHG allocation (Nordhaus 1994; Stern 2008). In essence, the DU model posits that rational human societies should discount future costs and benefits in favor of present costs and benefits.

Behavioral scientists, psychologists, and decision scientists, on the other hand, have empirically discredited the DU model (Frederick *et al.* 2002; Loewenstein 2003; Ariely 2009). Some have proposed an alternative hyperbolic discounting model, according to which people tend to be more impatient towards trade-offs involving earlier rewards than those involving later rewards. Yet others observe that the DU model cannot be salvaged by merely assuming a different—hyperbolic, for example—discount function. Rather, they argue, understanding inter-temporal choice behavior requires an account of several distinct motives that can vary greatly across decisions (Frederick *et al.* 2002).

There are significant ethical and algorithmic limitations when cross-scale value trade-offs are negotiated merely in terms of discounted utility or hyperbolic discounting models (Kelman 1981; Norton 1991, 1994; Page 1997; Sagoff 1998; Spash and Vatn 2006; Norton and Noonan 2007; Spash 2008). Instead of imposing positive discount rates through top-down environmental management and policy-making practices, behavioral, decision and system scientists argue that societal/stakeholder preferences must be elicited through a bottom-up deliberative type of processes. Further, an open-ended methodology must be deployed to elicit inter-temporal value preferences that permit decision-makers to display both positive and negative (or even non-linear) discount rates for different values when deciding about natural resource and other environmental management issues.

Drawing on this set of theoretical issues and debates, I posit the following hypothesis that is explored in this book in the context of mitigation and adaptation policy development for effectively responding to global climate change challenge:

(i) Negative Discounting Hypothesis: Hyper-discounting and positive discount rates do not accurately describe the decision behavior of policy actors in *all* natural resource management contexts; rather, negative discount rates for ecological and natural resource conservation values could also be observed in *some* management contexts.

Further, the measurement of trade-offs merely in terms of monetary costs and benefits may ignore other important social, ecological and political values, which are essential for context-sensitive management of natural resources but cannot be easily monetized or classified as costs and benefits (Norton and Steinemann 2001; Norton 2005; Spash and Vatn 2006; Spash 2008; McShane *et al.* 2011). Drawing on this insight, I posit a second hypothesis:

(ii) Value Pluralism Hypothesis: Ecological, social, political and other values could be accorded higher weights than economic values in *some* natural resource management contexts, both at present and future time scales.

We as social agents discount space and time. The space-time discounting poses challenging problems for environmental evaluation (and policy and decision analysis) at larger spatio-temporal scales such as manifested in the challenges posed by global climate change. I thus assert that we need a change in those environmental evaluation methods that assume that cross-scalar dynamics among individual agents and larger landscapes are not significant. We need an environmental evaluation methodology that incorporates heterogeneous spatial and temporal discounting by social agents and yet facilitates understanding of agent–environment interactions at multiple spatio-temporal scales.

Significant empirical research in social sciences has shown that we as social agents are faced with the problem of spatial and temporal discounting. For further details on spatial discounting, please see Perrings and Hannon (2001) and White (2004). For a detailed review about evidence on temporal discounting, please see Frederick *et al.* (2002) and Price (1993). This problem becomes more obvious and intractable when it comes to devising long-term plans and policies for protecting our environmental resources (e.g., atmosphere) at larger spatial scales. To accommodate this challenge, we need an environmental evaluation methodology that incorporates heterogeneous spatial and temporal discounting by social agents and yet facilitates understanding of agent–environment interactions at multiple spatio-temporal scales. The fact of positive interest rates empirically observed in all politico-economic systems of the world demonstrates one facet of positive temporal discounting. The economic agents/systems place higher value on objects closer in time and value decays as objects become distant in future. The higher the interest rate, the more quickly the objects become value-less.

Similar observations have been made in terms of spatial discounting. The closer the objects in space are to a social agent, the higher value is placed on the object. Consider the example of "not in my backyard" (NIMBY) phenomena observed in environmental research about the polluting agents. Or one might consider the problem of one community attaching lesser value to distant communities as compared to the neighboring communities (e.g., in environmental diplomacy or international trade models).

Let us make the concepts of spatial and temporal discounting formally clear for studying its implications on the long-term environmental evaluations. Assume that V_{ti} represents the (present) value of an environmental object E (e.g., environmental habitat, ecosystem, chunk of land, chunk of global atmosphere) for an agent k at time t and location i (with coordinates x, y, and z). Further, a spatio-temporal discount rate r_{ti} is defined that lets V_{ti} be mapped onto the value of the object E for agent A_k in some (past or) future time $t+u$ and/or some other location $i+j$, defined as $V_{t+u,i+j}$. Both u and j can be defined as discrete extensions in any space-time unit. The value mapping relationship for agent A_k is defined by equation 2.1:

$$V_{ti} = [1/(1+r_{ti})**u, j]*[V_{t+u,i+j}] \tag{2.1}$$

If r_{ti} is negative or positive integer, we have a simple example of a linear spatio-temporal discount function. The value of object is assumed to decay/discount linearly over space-time. In case r_{ti} is itself outcome of a non-linear differential function (or non-continuous function), then we have the case of a non-linear (non-continuous) spatio-temporal discount rate.

The most important feature to note in equation 2.1 is the nature of the patterns placed on the value of objects in future space-times when positive, null or negative discount rates are chosen by an individual agent. With positive discount rates, the net value of the object declines asymptotically and eventually becomes zero at some temporal scale. Only when discount factor of 0 percent is chosen, do we have a scenario that the value of object is assumed constant (i.e., same as at the present space-time) for all temporal and spatial scales. Conversely, when discount factor is negative, we observe that the value of the object tends to approach infinity as we extend the spatio-temporal horizon towards infinity. The assumption of negative discount rates would thus imply that the agent assumes the value of the object would be much higher at some future date or distant place as compared to the present space-time. Since empirical observations of social agents have shown that positive discount rates are virtually always employed in many decisions affecting our places and environments, it can be concluded that social agents assume a limited spatio-temporal horizon beyond which the (environmental) objects are assumed to have null value. This poses a fundamental problem with evaluating those objects (e.g., environments, places, habitats, etc.) that operate at much larger spatio-temporal scales such as the dynamics of global climate change operating at decadal, centurial and millennium timescales. By employing positive spatio-temporal discounting, social agents are prone to inherently under-valuing the objects which operate at larger spatio-temporal scales (e.g., longer than 250 years and/or farther away from agents' place of evaluation). Conversely, if agents employ negative spatio-temporal discounting, social agents would over-value the objects at larger spatio-temporal scales. Another possibility is that agents may have "patchy" spatio-temporal discounting, which might result in very obtuse valuations for environmental objects. Given this theoretical context, it must be noted that Stern (2008) came up with different damage functions in terms of global gross domestic product (GDP) loss with different social discount rate choices (1 percent to 10 percent). But why does human civilization have to choose a positive discount rate?

Climate change mitigation and the politics of temporal scale: "equitable" GHG emission allowances over multiple generations

Klinsky and Dowlatabadi (2009) synthesize relevant literature on conceptualization of justice in climate policy and, in particular, categorize 32 post-Kyoto policy architecture proposals on the criteria of their distribution rules and definition of the problem. Klinsky and Dowlatabadi (2009: 100–101) identified the lack of evaluative modeling as one of the major research gaps for comparing these policy architectures. Though some evaluative research that includes distributive justice is beginning to take place (Höhne *et al.* 2006; den Elzen *et al.* 2005; Rive *et al.* 2006), Klinsky and Dowlatabadi (2009: 101) assert that "it has focused almost exclusively on emission reductions and GDP." Further, Klinsky and Dowlatabadi (2009: 101) recommend the "work to improve our ability to index vulnerability could facilitate modeling of fair climate policies in which the burden includes impacts. This is an area where both the ethics and the policy communities could contribute."

While it should be an important long-term climate policy research goal to evaluate the climate change mitigation policy options contingent upon impact distribution and adaptation (which in turn is contingent upon the assessment of vulnerability and adaptive capacity), the proposed climate change mitigation policy architectures also need to be evaluated in terms of differentiated emission reduction burdens that are generated by different underlying decision heuristics. In other words, the debate on GHG emissions and development/GDP is not closed yet. The choice of a decision heuristic to allocate GHG emissions has direct implications on the emergent patterns of development and energy systems, which in turn also affects adaptive capacity and vulnerability patterns across the spectrum of developing and developed countries.

As argued in Chapter 1, the design of a post-Kyoto international climate governance regime will require a choice between grandfathering, or per capita or some other complex decision heuristic to allocate GHG emission allowances. The Kyoto Treaty employed a grandfathering decision heuristic. A grandfathering decision heuristic establishes a baseline year (1990 for Kyoto) and participating nations agree on binding targets and deadlines to reduce GHG emissions to a certain percentage below the baseline year, which is 5.2 percent below 1990 levels for certain developed, so-called Annex I, countries by 2008–2012 average period under the Kyoto Protocol (UNFCCC 1998: Appendix B). The grand-fathering decision heuristic has been severely criticized by the overwhelming majority of developing nations (including large GHG polluters such as China and India) for granting a "free ride" to the industrialized nations for the GHG emissions that were either emitted prior to the baseline year since the onset of the industrialized revolution in the 1750s or that are being emitted well above global per capita averages since the baseline year. Some of these GHG emissions will stay in the atmosphere for hundreds and thousands of years and, representatives of developing countries argue, any fair GHG emission reduction plan must take into account these GHG emissions that were an unintended consequence of the industrialization process that brought overwhelming development to the industrialized world.

The politics of scale are manifested in the choice of a "baseline" year in using grandfathering type of decision heuristics for setting up GHG mitigation targets. Even a slight shift from 1990 to 1988 could have resulted in the absence of so-called "hot air" emissions for Russia and other eastern European Annex I countries in the KP. Similarly, the problem of historical GHG emissions could not be just pushed under the carpet, primarily because many GHG emissions are cumulative at the millennium timescale while GHG budgets are being negotiated at centurial timescale. Consider the differences in GHG emissions shown for industrialized countries in Figures 2.1 and 2.2 from 1990 to 2010 as reported to UNFCCC. In Figure 2.1, only annualized total GHGs are shown, while in Figure 2.2, cumulative GHGs are shown. Most important is to note the difference in the scale of Figures 2.1 and Figure 2.2: While the *scale* of Figure 2.1 runs from 5000 to 20,000 terra grams of CO_2 equivalent, the *scale* of Figure 2.2 runs from 100,000 to 300,000 terra grams of CO_2 equivalent. Due to the *accumulative* nature of GHGs in the atmosphere, Figure 2.2 should be used as the standard method to communicate and discuss the mitigation targets by UNFCCC and other international organizations; however, due to the politics of scale, annualized GHG-based figures (e.g., 2.1) are more widely used to compare and set GHG reduction targets. From Figure 2.2, we can infer that the United States is responsible for emitting more than 100 billion tons of CO_2 equivalent GHGs in the atmosphere between 1990 and 2010, which is strikingly different when it is typically stated that the United States emits about six to seven billion tons of CO_2 equivalent GHGs every year, as shown in Figure 2.1! Furthermore, even more importantly, the grandfathering based baseline approach initiated in KP uses annualized GHGs shown in Figure 2.1, which is again misleading and completely ridden with the politics of scale. Due to the accumulative nature of GHGs, cumulative GHG based approaches, as shown in Figure 2.2, must be utilized when assigning GHG entitlements and GHG budgets in a post-Kyoto climate governance regime.

The "grandfathering" decision heuristic based upon annualized GHG emissions is not an equitable option for a post-Kyoto climate governance regime. Instead of the grandfathering decision heuristic, *cumulative* GHG emission caps on a per capita basis will be more "equitable." The cumulative GHG per capita

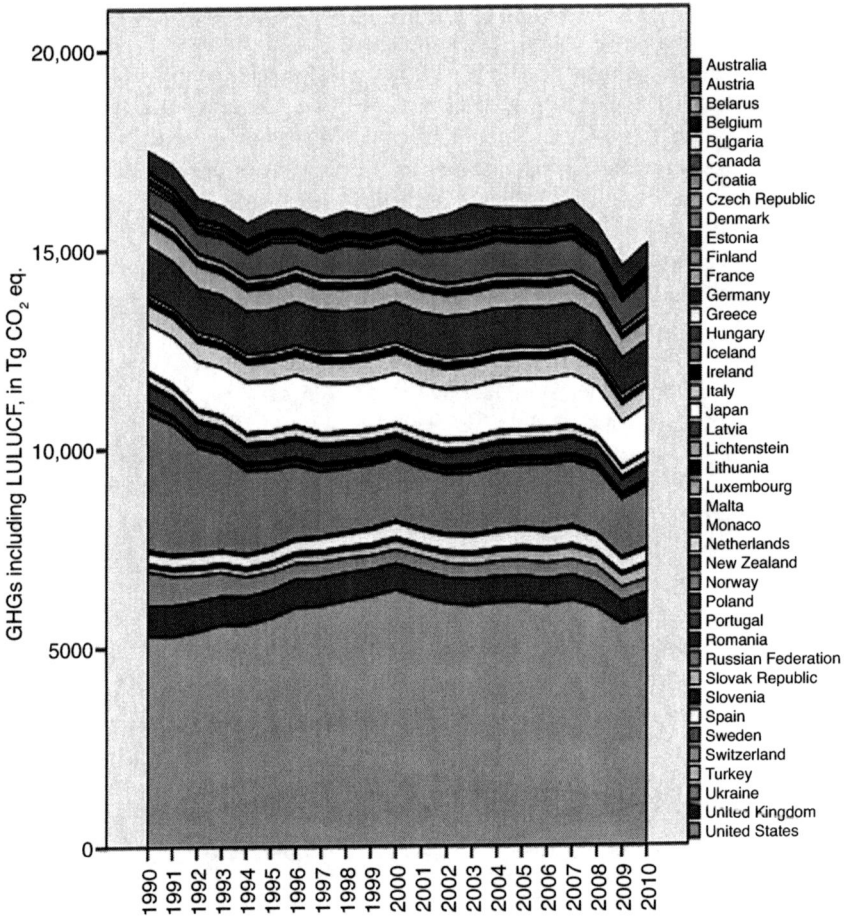

Figure 2.1 Annualized GHG emissions for developed countries (1990–2010) (data source: UNFCCC).

decision heuristic is based upon the ethical principle that all people on this planet must be treated as "equals" with an equal right to emit equal emissions. This GHG per capita decision heuristic will also preserve the right of development for developing countries. Further, these cumulative GHG per capita targets must be estimated by properly considering a "fair" time scale that includes both historical and future GHG emissions. Perhaps, the best approach will be to agree on a target GHG concentration level by a specific deadline, such as 350 ppm by 2050, 450 ppm by 2100, or 350 ppm by 2150. While the choice of a target quantity and target deadline will entail complex politics of scale, it will enable fairly straightforward estimates of cumulative GHGs since the onset of industrial revolution for each nation state. This approach could lead to a global "GHG budget" as well

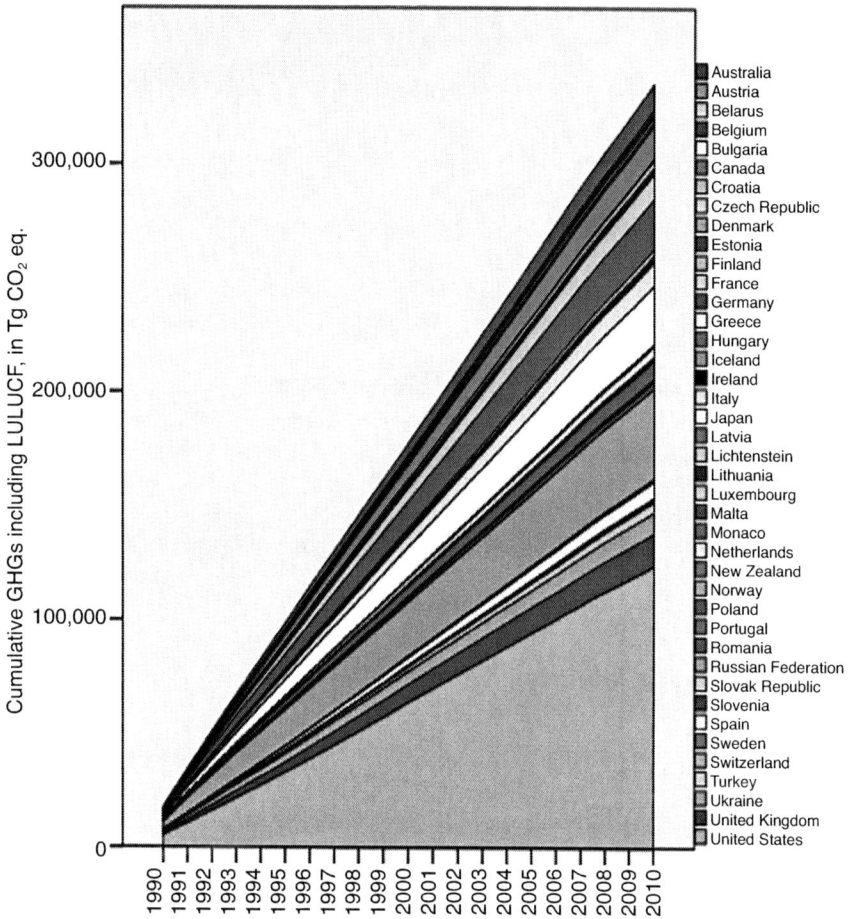

Figure 2.2 Cumulative GHG emissions for developed countries (1990–2010) (data source: UNFCCC).

as national scale "GHG budgets." While there is large GHG per capita variance within and across countries, average cumulative GHG per capita targets for each country established through such a mechanism could provide a "fair" solution without requiring any settlement on discount rates (e.g., Stern review controversies on the choice of discount rates). The fundamental question is whether such cumulative GHG per capita approach derived from cumulative GHG budgets based upon agreed upon global targets would be acceptable to the industrialized countries, especially those countries which have relatively very high GHG per capita emissions, as shown in Figures 2.3 and 2.4 for the industrialized countries from 1990 to 2010. A related question is, if the industrialized countries are not willing to accept binding commitments along these lines, is there a case for imposing trade sanctions and other forms of criminal charges against these countries for causing global climate change. Such questions are not even included in the agenda of UNFCCC negotiations because there is an underlying politics of scale here, played by rich industrialized countries against the poor developing countries.

The leading GHG per capita emitting developed countries, especially the United States, Canada and Australia, as shown in Figures 2.3 and 2.4, have consistently shown their consistent disagreement with the per capita decision heuristic. These developed countries argue that per capita decision heuristic will not only result in more stringent caps on the developed nations with higher historical GHG emissions, but it will also allow for larger populations of the developing countries. Implicitly, these developed countries are thus arguing that their citizens have higher rights to emit GHGs because they are more equal than others! The ethical dilemma surrounding the choice of one or another decision heuristic will require a resolution for designing an equitable post-Kyoto governance regime. This ethical dilemma is further compounded by the choice of temporal and spatial scales on which GHG per capita budgets will be estimated for such a policy arrangement. I argue that such a policy arrangement will not only directly reduce GHG emission growth rates in industrialized countries, but it could also potentially have a very large effect in changing the development pathways for

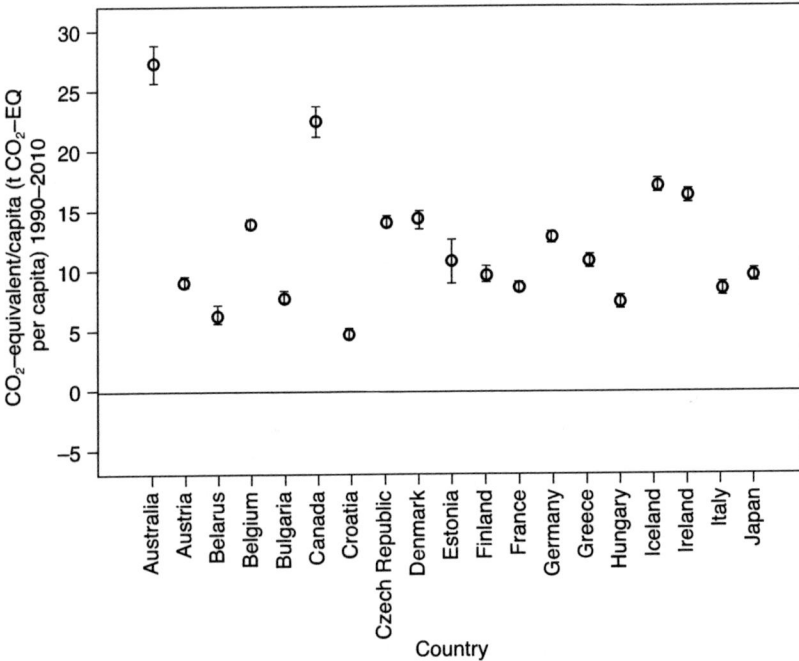

Figure 2.3 GHG per capita emissions for developed countries (A to J) (data source: UNFCCC and World Bank).

developing countries. They could "leapfrog" from very GHG intensive to less GHG intensive development pathways.

Let us consider the case of developing countries, which I will define as countries whose annual GDP is currently below $20,000/capita. These countries have contributed a very small percentage of cumulative greenhouse gas (GHG) emissions since the dawn of the industrial era as compared with rich industrialised countries. The developing countries face stark policy choices in their energy, transportation, agriculture and forestry sectors in the twenty-first century: they can either, as shown in Figure 2.5, follow the carbon-intensive development path followed by rich industrialised countries ("business-as-usual"—BAU—scenario) or "leapfrog" to a sustainable development (SD) scenario. There is a continuum of possible future development scenarios that is available to developing country policy-makers; however, they must realize that, in retrospect, there is only one development path that would have been eventually followed by these countries by the end of the twenty-first century. It could be something worse than the BAU path (e.g., Liquified Coal scenario), BAU path, an SD path, or something in between. Given large system lags and huge upfront infrastructure build up costs, energy, transportation, agriculture and forest systems require medium- to

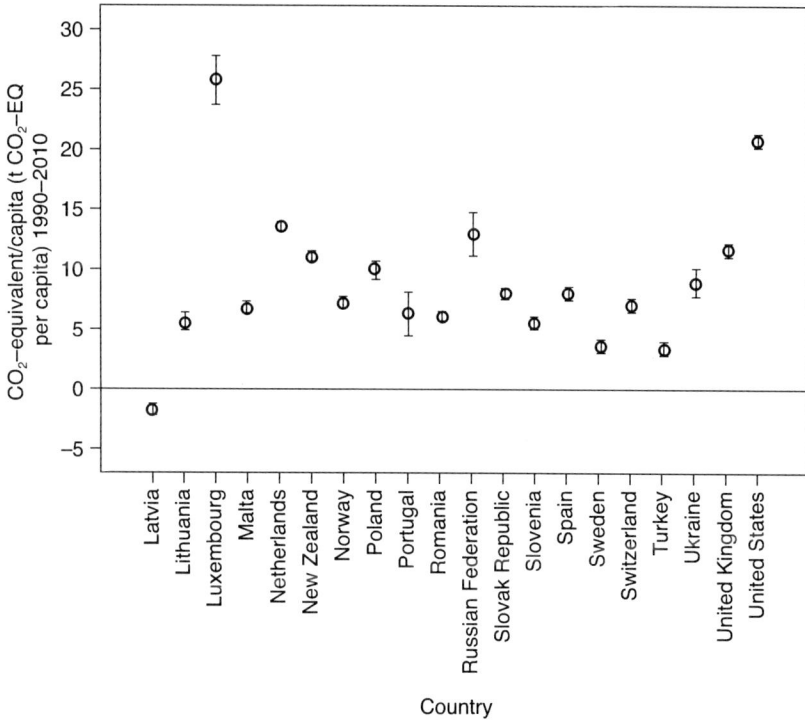

Figure 2.4 GHG per capita emissions for developed countries (L to U) (data source: UNFCCC and World Bank).

long-term planning, for which reason a fifty to 100 year time span is more appropriate to consider than a five to ten year one.

There are three related questions that will need to be addressed by the policy-makers in the developing countries if GHG per capita entitlements are agreed upon at a global scale:

1   What are the trade-offs between leapfrogging to a SD path versus sticking with a BAU path?
2   How can system-wide carbon-intensity indicators vis-à-vis Per Capita Income enable developing countries to keep track of SD versus BAU paths?
3   Which policy changes in energy, transportation, agriculture and forestry sectors are required to shift developing countries from a BAU to an SD path?

The relatively higher upfront cost of investment in renewable energy systems, which have lower carbon intensity, is the most debilitating barrier to shift from BAU to SD paths. However, what the policy-makers must realize is the cost of inaction: if BAU continues, they might be able to save some money for the current generations, but their future generations in the next fifty to 100 years will have to pay relatively higher costs—effects such as poor air quality from fossil fuel consumption in the energy and transportation sectors, and adverse climate change impacts such as intensified droughts, floods, typhoons, etc. This is another example of politics of temporal scale that the policy-makers and citizens in developing countries will need to be engaged in as GHG entitlements are decided through international governance mechanisms.

Another tactical set of problems for developing countries will be oil and energy security challenges. Energy dependence, especially on petroleum and natural gas, can potentially trigger strategic wars under a BAU scenario. Triangulating these factors, a switch from a BAU to a SD path will be potentially of much higher long-term value. So, a trade-off exists between huge upfront but relatively lower costs borne by the current generation of developing countries versus very high costs borne by future generations in these countries.

Given the inertia in policy systems, especially in energy, transportation, agriculture and forestry sectors, it will be a huge challenge for policy-makers to change the status quo of BAU. In fact, some proponents of unfettered growth might argue that future generations of developing countries will be rich enough to deal with environmental impacts, hence we should now focus on the most cost-effective development path, which due to perverted cost measurements turns out to be carbon-intensive path, as shown in Figure 2.5, for a majority of the rich nations so far.

Figure 2.5 shows CO_2 (tons) per capita against GDP per capita for all countries of the world in a panel data set from 1971 to 2004 constructed from IEA (2007). The BAU Carbon-intensive Development path in Figure 2.5 captures the linear relationship between the CO_2 per capita and income (GDP per capita) of about 120 countries/regions represented in the database. For example, a country like the United States has attained very high per capita GDP (~$35,000 per year), but due to its carbon-intensive path, it also has very high CO_2 per capita (~25 tonnes per year).

For developing countries to increase their per capita income to U.S. levels, they will require a massive energy input in their economic systems. Figure 2.6 shows energy per capita plotted against energy per GDP over time for the United States, four South Asian countries and the world average. The United States has maintained its high per capita income with an energy input of about eight tons of oil equivalent (TOE) per capita. In contrast, the global average is about 2 TOE per capita. The South Asian countries are, so far, well below 1 TOE per capita but they are slowly increasing their per capita energy consumption. In order to switch away from a BAU towards a SD path, the additional energy per capita for South Asian countries must come from renewable energy sources to keep the carbon intensity lower. One encouraging trend is that energy intensity (i.e., energy per GDP), as shown in Figure 2.6, is decreasing over time for South Asian countries. This trend could be expedited under the SD path if GHG per capita entitlements are agreed upon.

Carbon intensity, measured as CO_2 per unit of energy, is one of the most critical indicators that will differentiate BAU versus SD paths over time. Figure 2.7

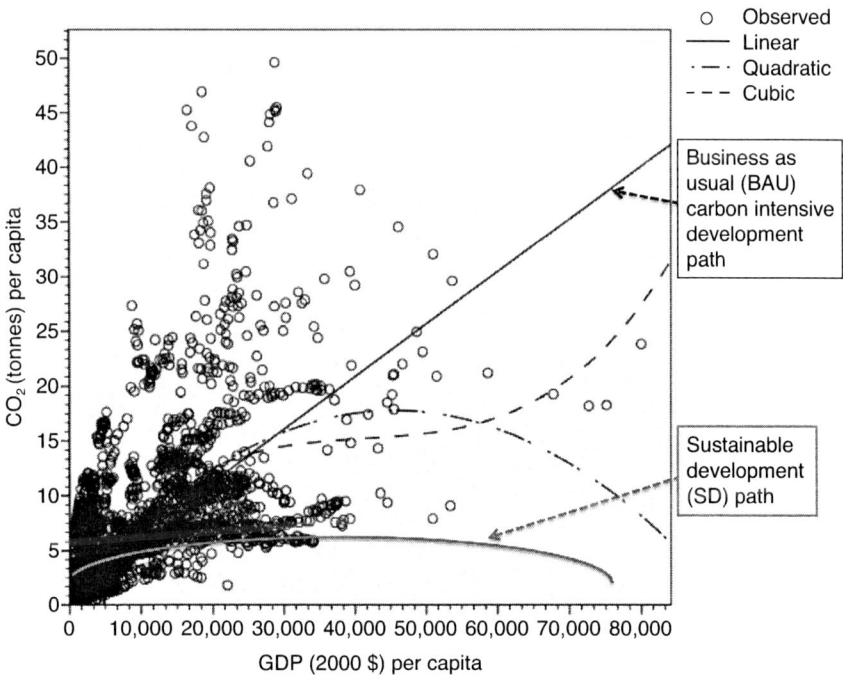

Figure 2.5 CO_2 per capita plotted against GDP per capita for 120 countries (data source: IEA 2007).

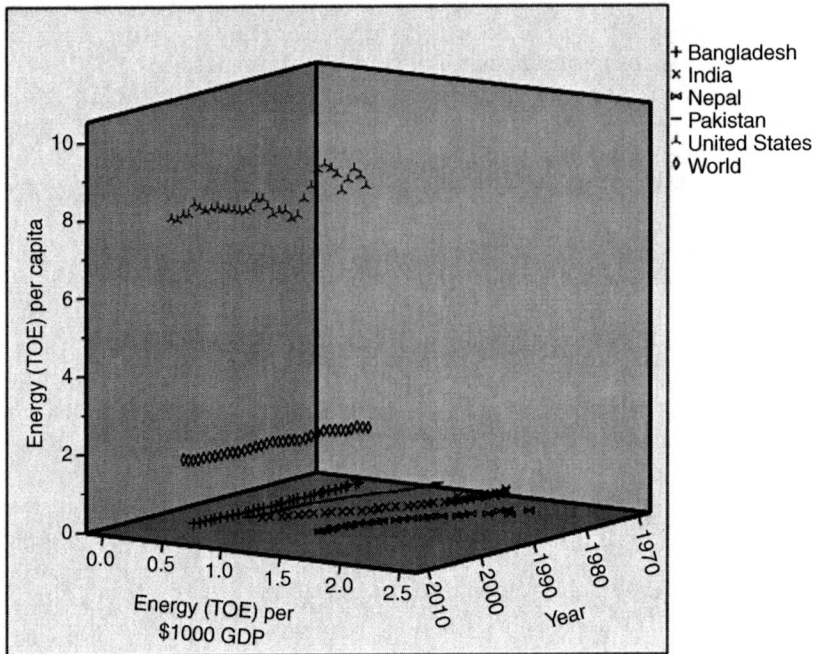

Figure 2.6 Energy per capita vs. energy per GDP for USA, India, Pakistan, Bangladesh, and Nepal (data source: IEA 2007).

shows carbon intensity, measured as tons of CO_2 per TOE of energy, against CO_2 per GDP over time for four South Asian countries, United States and world average. While United States and world average for carbon intensity has been much higher between 1970 and 2004, their trend is towards decreasing carbon intensity over time. In contrast, South Asian countries (with the exception of Nepal, courtesy their hydropower) have an upward trend of increasing carbon intensity during the same period. Under BAU, these upward trends of carbon intensity will continue, while under a SD path, South Asian countries will be able to level off and then decrease the carbon intensity of their economies, while at the same time increasing their per capita income. In other words, leapfrogging to a SD path will happen if developing country policy-makers can reverse the trend of increasing carbon intensity. While CO_2 per GDP, as also shown in Figure 2.7, is decreasing, which is an encouraging sign, the real challenge is to arrest the growth of carbon intensity of energy consumption in developing countries. This can be accomplished by a series of policy changes at these early stages of development.

As the economies of the developing countries gather momentum to increase their per capita income during the twenty-first century, an early and long-term

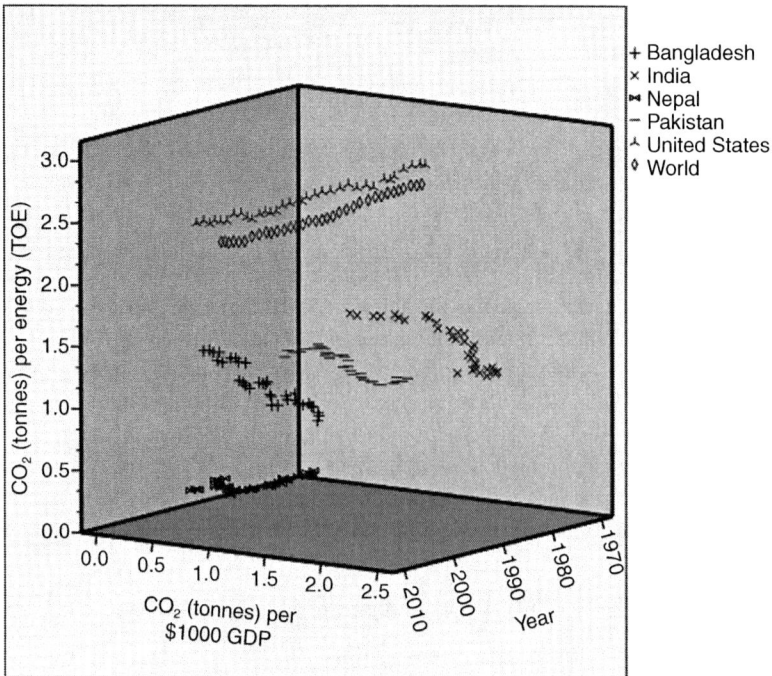

Figure 2.7 CO_2 per TOE vs. CO_2 per GDP (data source: IEA 2007).

action by the policy-makers in these countries can enable a switch from a BAU to a SD path. Major policy changes for this switch will directly impact energy, transportation, agriculture, and forestry sectors, but indirectly every sector of the economy will be affected due to the proposed policy changes. In the energy sector, carbon-intensive energy inputs (especially coal, natural gas, petroleum) should be taxed and the revenues from these taxes should be spent on developing renewable energy systems (solar, wind, micro-hydro, biomass, geothermal, ocean). This will prevent economic actors and firms from latching on to carbon-intensive energy systems and will help promote renewable technologies. While this policy will change the winners and losers in these societies, the overall cost of these changes in the long run (50 to 100 years) will not adversely affect income growth.

The transport sector's addiction to petroleum and diesel must be avoided through a series of explicit policy interventions to improve land use planning, enhance public transit systems and promote lightweight fuel efficient hybrid or electric vehicles. More than 40 policy strategies for reducing the carbon intensity of the transportation sector are shown in Figure 2.8.

For analytical convenience, in Figure 2.8, the 44 policy options are organized under two major categories: (a) reducing CO_2 per vehicle mile traveled (VMT), and (b) reducing VMT/year. Reducing CO_2 per VMT, in general, refers to the broad category of policy options that improve efficiency while reducing VMT/year refers to the set of policy options that improve energy conservation by providing viable alternatives to reduce VMT from gasoline driven vehicles. In Figure 2.8, CO_2 per VMT reduction policy options are sub-categorized for passenger vehicles (14 options) and freight operations (seven options).

The policy option of low GHG tailpipe standards is contested by U.S. auto makers, but if implemented it will potentially result in more fuel efficient and lower GHG/mile fleet averages, which will change the system-wide GHG impact of transportation activities. Feebates will similarly incentivize consumers to buy

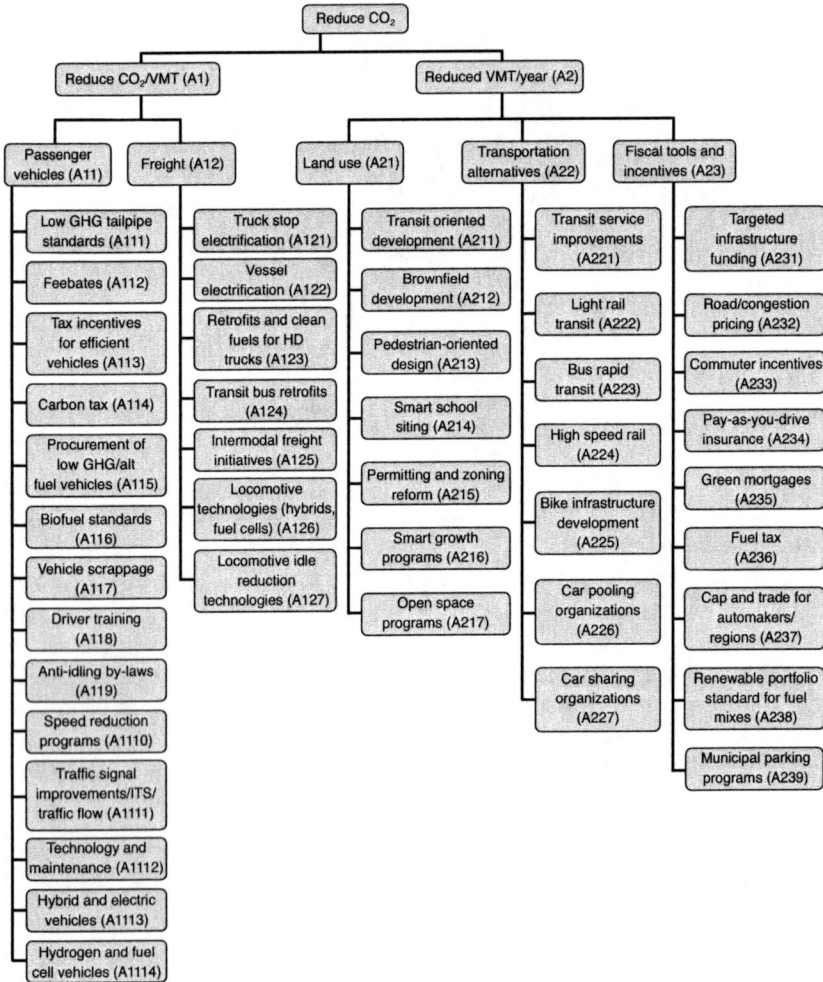

Figure 2.8 A snapshot of forty-four policy options to reduce GHG emissions from transportation sector.

more fuel efficient vehicles. Negative feebates against inefficient vehicles can also be instituted. Carbon tax will potentially also result in lower GHG/mile from transportation activities. Procurement of low/alternative fuel vehicles will reduce GHG/mile, as will biofuel standards. Vehicle scrappage can lead to cleaner and more fuel efficient vehicular fleets active on the roads. Driver training can improve their acceleration/deceleration practices and lead to lower GHG/mile. Laws against idling can reduce GHGs emitted for no obvious transportation activity. Introduction of speed reduction laws and intelligent transportation systems in all major and minor traffic arteries can reduce GHG/mile. Proper maintenance of vehicles, such as proper tire inflation, can also reduce GHG/mile. Introduction of more hybrid, electric, or fuel cell driven vehicles can also cause total life-cycle reductions in GHG/mile. In a nutshell, all of the fourteen policies listed under A11 in Figure 2.8 have the potential to lower GHG/mile for all passenger vehicle transportation activities. Freight options under A12 similarly apply to lower GHG/mile from freight-related transportation activities. Similarly, VMT per year reduction policies are sub-categorized as land use (seven options), transportation alternatives (seven options) and fiscal tools and incentives (nine options). Transit oriented developments can potentially shift more and more agents to take public transportation as their preferred modal choice, thus potentially reduce VMT per year. Brownfield development can reverse suburbanization trends and potentially lead to more urbanization and reduce transportation activity caused by suburban developments. Pedestrian oriented designs can be used to encourage walking as the preferred transportation activity and, implicitly, to reduce VMT per year. Smart growth programs aim at minimizing the need for transportation activities and provide easier access to public transport systems, thus overall reducing VMT per year. In a nutshell, all the policy options listed under A2 can reduce VMT per year.

The ultimate goal of all the policy options shown in Figure 2.8 is to stimulate a change in the current state of transportation systems, either towards reducing GHG/VMT or VMT per year. Any change in the current state of transportation systems through policy interventions will inevitably cause a change in the transportation activities (for both VMT per year and GHG/VMT) of individuals, firms and other entities. However, noticeably, some policy interventions can be implemented at much shorter time scales (e.g., carbon taxes), while others will require protracted planning over longer time periods (e.g., land use changes). Such scale issues engender politics of scale when comparing and choosing such GHG mitigation policies within and across transportation, energy and agriculture sectors.

The agriculture sector is typically a major producer of methane and nitrous oxide (two of the very potent GHGs), but technological agriculture could also produce relatively high levels of CO_2 emissions. Adverse impacts of the so-called "green revolution" on the productivity of lands in many developing countries have already started to appear in some areas. Explicit policies to promote organic farms, integrated pest management strategies and less carbon-intensive farming techniques (through decentralised windmills and solar applications) will reduce overall GHG emissions from the agriculture sector in developing countries. Finally, the conservation of old-growth forests will allow natural carbon sequestration, while growth of new forests to produce renewable biomass materials, such as cellulosic ethanol for transportation systems, can reduce overall carbon intensity. There is no single silver policy to switch from BAU to SD paths, but the overall effect of each of these individual policies can be much greater than the sum of individual policy shifts in energy, transportation, agriculture and forestry sectors to wean developing countries off a BAU path and lock them into a SD path without any caps on economic growth.

While technological possibilities exists to "leapfrog" developing countries from a carbon-intensive development path to a renewable energy driven less carbon-intensive path, the politics of temporal discounting applied by typical policy-makers in developing countries, and developed countries for that matter as well, lead to the appearance of carbon-intensive development paths as more "cost effective" and "efficient." The politics of temporal scale can be unlocked by both expanding the temporal scale on which different technologies and GHG mitigation policies are evaluated as well as appropriately using positive, neutral or negative discount rates.

Spatial discounting and intra-generational allocation problems

The tension between conservation and development objectives across the globe is a function of many complex issues (Hirsch *et al.* 2011; McShane *et al.* 2011), one of which concerns how to trade off pluralistic values associated with anthropogenic environmental change occurring at multiple scales of space and time. Recently, a considerable amount of research and scholarship has been devoted to understanding the cross-scale trade-offs that ensue from the management of social-ecological systems at multiple levels of social organization (e.g., Berkes 2002, 2006; Adger *et al.* 2005; Brown and Purcell 2005; Lebel *et al.* 2005; Rodríguez *et al.* 2006; Silver 2008). Another method to estimate cross-scale trade-offs in the management of social-ecological systems concerns the politics of spatial scale, which involves an explicit focus on the ways in which powerful actors at larger scales of social and spatial organization influence the policies and management of social-ecological systems at relatively smaller scales. While literature on the politics of spatial scale in political economy and political geography (Smith 1992, 1993, 1995; Jonas 1994; Agnew 1997; Delaney and Leitner 1997; Swyngedouw1997a, b, c, 2000; Howitt 1998; Marston 2000; Brenner 2001; Escobar 2001), and more recently political ecology (Brown and Purcell 2005; Cash *et al.* 2006), has focused on assessing the strategies pursued by individuals or groups across different spatial levels of social organization to achieve a particular agenda, this book incorporates the basic insights put forward by theorists of the politics of scale with the goal of quantifying differences in the valuation of alternate climate policy architectures across multiple spatial scales of social organization.

A central theoretical argument is that alternate climate policy architectures, each of which are associated with different mixes of development and conservation goals, lead to the emergence of asymmetric distributions of value for different social organizational groups across spatial scales. Management of social-ecological systems is thus a dynamic interplay of politics of scale, creating and recreating winners and losers at multiple spatial scales with the implementation of different mixes of conservation and development policies.

Tropical deforestation contributes anywhere from 17 to 25 percent of global GHGs, yet local communities in tropical countries appear to have negative attitudes towards forest conservation for a variety of reasons, such as the restrictions imposed by the protected area authorities which deny local communities access and user rights on natural resources for agriculture and livestock production. The relocation/eviction of villagers, according to many local communities, results in the loss of arable land, settlements and livestock during the eviction process, and the eviction process interferes with their cultural and traditional beliefs and rituals. Such issues of local communities represent the larger conservation-development conflict for the management of social-ecological systems at the global scale. Many of these conflicts have arisen in the post-colonial context of historical grievances and contestations of benefits sharing among social organizations at different spatial scales. The conservation community has for decades struggled with, and argued over, the spatial distribution of costs and benefits that accrue from natural resource conservation practices such as the eviction of local communities from the expansion of national parks in Africa (Adams and McShane 1996; Songorwa 1999; Agrawal and Redford 2006; Brockington *et al.* 2006; Wilkie *et al.*, 2010). The costs of conservation actions (e.g., evictions of local communities) are imposed on local and indigenous communities while national and international communities enjoy the benefits of ecological conservation (e.g., less GHGs in the atmosphere). Conversely, political economists have argued that the environmental costs of open-ended development actions (i.e., extinction of animals and plants) are also typically imposed on local and indigenous communities while national and international communities enjoy the benefits of globalization and economic development. This leads to a more general hypothesis that larger scale social organizations derive higher net benefits from pure conservation or pure development management options while smaller scale social organizations derive lower net benefits from such "pure" management options. The politics of spatial scale in promoting tropical forest conservation has been the subject of numerous other types of research concerning conservation and its relationship to rural communities, and the numerous interventions attempting to address such discrepancies (Agrawal and Gibson 1999; Hulme and Murphree 2001; Brokington and Schmidt-Soltau 2004; Wells and McShane 2004; Sunderland *et al.* 2007).

The politics of spatial scale observed in tropical countries is symptomatic of many similar conservation type management scenarios in both developing and developed countries. Transparent explication of spatial distribution of value trade-offs (e.g., Zia *et al.* 2011) enables an integrative analysis to identify and discuss alternate management options that could potentially balance the current politics of scale back towards local level social organizations. When policies are extreme versions of open-ended development, or on the opposite end, an extreme version of regulated conservation, larger scale social organizations appear to derive higher value while smaller scale social organizations derive lesser value. Since both explicit and implicit forms of power are available to social organizations at larger scales, we observe this type of politics of scale in the management of many different kinds of social-ecological systems. In contrast, the politics of scale could be potentially mediated by designing non-extreme policies that balance conservation and development goals. Further, quantification of cross-scale value distributions could also provide insight regarding the appropriate magnitude and direction of payments for ecosystem services and other compensatory mechanisms from international and national level to local level social organizations in situations where balanced or mixed management approaches are not feasible. The quantification of such cross-scale value trade-offs could provide useful information in the future for the design of policy mechanisms that transfer benefits from international and national to local levels of social organization and, hopefully, reverse the current course of politics of spatial scale in managing social-ecological systems. It is not only in climate change mitigation policy development that the politics of scale are observed, but climate change adaptation is also afflicted with this.

Climate change adaptation and the politics of "risk zoning"

To govern the risk from natural hazards in the face of complex global environmental and societal dynamics, various types of publicly sponsored risk insurance mechanisms have been proposed as proactive policy options (MacKellar *et al.*, 1999; Amendola *et al.*, 2000; Linnerooth-Bayer *et al.*, 2003; Linnerooth-Bayer and Mechler 2007, 2008). Many of these risk insurance mechanisms implicitly or explicitly use the concept of "risk zones" to establish a legal basis for governmental intervention. Broadly, a risk zone may be defined as a three-dimensional spatial frame that is exposed to an exogenous or endogenous hazard over time. Clearly, the politics of (spatial) scale can be observed in those climate change adaptation policies that require the determination of risk zones, such as the institutionalization of public, private, or public–private risk insurance mechanisms, vulnerability reduction programs and/or resilience enhancement initiatives.

Promoting hazard-based risk insurance schemes in designated risk zones, such as the U.S. National Flood Insurance Program (USNFIP), can be used to mitigate, redistribute or absorb the risk from natural hazards and design "resilient" communities. I assume that the higher the capacity of a community to absorb risk from natural hazards, the more "resilient" is that community. The development and implementation of drought and hurricane insurance policies, and long-term land use planning for adaptation to climate change, also require the estimation and designation of risk zones (Smith and Lenhart 1996; Adger *et al.* 2005a; Botzen *et al.* 2009).

Different governments govern the risk from natural hazards differently (Glantz 2003). Some governments are more proactive, while others are more reactive. Some governments break down the hazards in different categories to deal with each separately, while others integrate the hazards in a holistic governance mechanism to deal with all types of hazards. Some governments leave it to the insurance markets to mitigate the risk, while others regulate or even subsidize the insurance markets to redistribute the risk. In some cases, governments assume the role of insurance companies. Further, regulation of development patterns and building codes also varies broadly across the spectrum of governments from local to national and international levels. The prevalence of variation in risk governance mechanisms across different governments, both horizontally and vertically, provides a rich empirical setting for climate change policy analysts to compare the effectiveness of risk governance mechanisms. The equity, efficiency and other impacts of risk governance mechanisms must also be assessed on a regular basis.

The periodic analysis of differential risk governance mechanisms has the potential to transfer good practices, learnable lessons and effective mitigation strategies from one community to another. There is a huge potential to learn from one government's experience before applying similar risk governance policies in other socio-economic and technological contexts. This policy and governance learning process, however, is very dynamic and context-specific and cannot be seamlessly transferred from one place to another.

To investigate various risk governance issues from the perspective of the developing countries, I organized an international workshop in Islamabad, Pakistan in 2005. It was during the group discussions in this workshop as well as interviews with different stakeholders participating in the workshop that I found the issue of designating "risk zones" to be a salient problem for designing risk insurance policies aimed at mitigating risk from natural disasters. More details about this are in Zia and Glantz (2012). In particular, I was struck by many generalizable "wicked" challenges that were voiced by workshop participants for designating risk zones. I designate these issues as "wicked" because they appear to contain similar characteristics that were defined by Churchman (1967) and Rittel and Webber (1973) for "benign" versus "wicked" policy and planning problems such as there is no definitive formulation of a wicked problem; wicked problems have no stopping rule; and solutions to wicked problems are not true-or-false, but good-or-bad. Ackoff (1974) characterized these problems as "policy messes." There is growing body of literature on wicked problems in public policy and public administration (e.g., see Roberts 2000; Weber and Khademian 2008). In this context, qualitative and interpretative processing of workshop and interview data led to the identification of the following ten "wicked" challenges that were raised for the designation of risk zones to introduce public policy interventions as part of risk governance strategies:

1 Risk thresholds: what cut-point criteria are used to designate risk zones?
2 Land value: what are the effects on the land value when designated as risk zones?

3   Damage reduction: do publicly funded or micro-insurance programs in designated risk zones reduce damages?
4   Land-use planning: do designated risk zones contradict or complement the current local or provincial level land-use planning and zoning practices?
5   Forecast uncertainty: how is the forecast uncertainty incorporated in the risk premiums for national or micro-insurance programs?
6   Costs of accurate maps: what levels of investments are needed to continuously update risk zoning maps? What are the trade-offs between accuracy and costs?
7   Modifiable area unit problem (MAUP): how should individual or community level actuarial risk be determined in risk zones through aggregate level data?
8   Winners and losers: who are the winners and losers when public or micro-insurance programs are funded in designated risk zones?
9   Single versus multiple hazards: should risk zones be established for each hazard separately or combined with other hazards? Further, should public insurance and micro-insurance programs be oriented to mitigate single or multiple hazards?
10   Cross-jurisdictional administrative boundaries: how should public and micro-insurance programs be designed and managed when designated risk zones cut across established administrative boundaries?

These ten "wicked" design challenges point to the role of broader societal dynamics and policies in differential risk exposure of communities in developing and developed countries. Comfort *et al.* (1999) expressed this concisely:

> there is a widespread failure to recognize and address connections between changes in land use, settlement policies, population distributions and the accompanying degradation of habitats on the one hand and dramatically increased levels of hazard exposure and vulnerability on the other. We propose that human vulnerability—those circumstances that place people at risk while reducing their means of response or denying them available protection—becomes an integral concern in the development and evaluation of disaster policies.
>
> (Comfort *et al.* 1999: 39)

While policies, such as USNFIP and experimental micro-insurance programs in countries like India, assume that risk zones can be designated through some rational or expert-based system, Zia and Glantz (2012) argue that the designation of risk zones is also a political, social and value-laden process. Further, expert-based systems can be gamed to serve the political and normative goals of more powerful decision-makers, often at the cost of vulnerable populations, for whom risk governance mechanisms apparently purport to be designed. Based on these arguments, Zia and Glantz (2012) recommend that the designation of risk zones should be a more transparent and participatory process and vulnerable groups should be especially empowered to participate in the risk-zoning related deliberations. To facilitate these deliberations among scientists, policy-makers and affected stakeholders, Zia and Glantz (2012) undertook an in-depth analysis of these ten design challenges from the empirical perspective of flood insurance programs, such as USNFIP and other similar programmatic experiences, and derived key deliberative heuristics that could be used by policy-makers in both developed and developing countries for setting up participatory risk governance regimes. All of these ten deliberative heuristics are derived to propose a participatory policy development governance process that could be used in developing countries, where most of the adverse impacts of global climate change are expected to take place at least in the first half of twenty-first century. However, it will be a great fallacy to assume that developed countries will be spared the wrath of unmitigated climate change, hence the politics of spatial scale is a worldwide phenomena when viewed in the context of adaptation policy development.

For synthesizing evaluation literature on USNFIP and other such programs vis-à-vis challenges of designating risk zones, Zia and Glantz (2012) searched in the Google Scholar, ABI/INFORM global (proquest); Academic Onefile; and Academic Search Premier to select a sample of published papers that address *"flood insurance programs."* After an initial review of more than 120 papers, they sampled 56 papers. These papers were shortlisted from a larger pool based on their relevance to the study goals, i.e., whether they addressed any of the ten wicked challenges in publicly sponsored flood insurance mechanisms and, above all, whether they were peer reviewed. These fifty-six papers represent a broad array of disciplines ranging from economics, planning, geography, disaster management, risk management, and law to policy. Since a majority of these 56 "peer reviewed" studies are focused on evaluating USNFIP, the sampled papers do not adequately represent publicly funded flood insurance programs in non-English speaking countries (e.g., France) or privately funded flood insurance programs (e.g., England), which is a limitation of this study. Following Boruch and Petrosino's (2004) methodology of implementing research syntheses, Zia and Glantz (2012) synthesized the findings from fifty-six papers on ten wicked challenges for designating risk zones that were identified during Islamabad workshop and interviews. Each of these wicked challenges pertain to the manifestation of politics of spatial scale, as explained below.

Risk thresholds

Setting risk thresholds for designating risk zones is one of the most contested scientific and socio-political problems. The flood insurance program in the United States defines the Special Flood Hazard Area (SFHA) as the area of land that would be inundated by a flood having a 1 percent chance of occurring in any given year (also referred to as the base-flood or the 100-year flood). These areas are delineated on Flood Insurance Rate Maps (FIRMs), which are generated in the United States by FEMA. More information is available at www.fema.gov/hazard/map/firm.shtm. Flood frequency analysis is estimated through statistical models for identification of those communities at risk of flooding.

The criterion of determining 100-year floods is an example of threshold-based decision criteria, which means that changes in thresholds will result in sensitive variation of designated flood risk zones. The 1 percent-chance flood is an arbitrary criterion and its estimate is uncertain, particularly with climatic change. Uncertainty in the estimate of the 1 percent-chance flood could mean that residents outside of the SFHA actually live in an area where flood risk is higher than a probability of 1 percent in any year. Green and Petal (2008) consider the flood zones in the UK inadequate because of the inaccuracies in predicting flooding due in part to urbanization, climate change, and the lack of historical records. Further, neglect of coastal erosion (Kriesel and Landry 2004) or overestimation of levee strength (McKenzie and Levendis 2010) could also cause such misperceptions among the populations living outside designated flood zones. Young (2008) argues that remapping of 500-year floodplain maps could potentially extend mandatory coverage. Zahran *et al.* (2009) propose development of Community Rating Systems (CRS) to designate insurance premiums, with discounted premiums for those communities that demonstrate preparedness for the flood risk.

Local governments can potentially adopt more stringent criteria than those required by the USNFIP. One suggested adaptation to climatic uncertainty could be to regulate floodplains with a lower probability of flooding than 1 percent per year, but regulation outside of the SFHA is expected to be unpopular politically due to the perceived adverse economic effects on property owners. In the Netherlands, as discussed by Botzen *et al.* (2009), the "dike rings," governed by the "Water Embankment Act of 1996," have a safety goal of protecting against 10,000-year or 0.0001 percent chance of flooding. Coastal Building Zones in Florida represent special criteria mandated by local communities to deal with wind, beach erosion and flooding simultaneously (Dehring 2006a, b). In summary, Zia and Glantz (2012) found that the predefined 100-year, 500-year, or 10,000-year criteria cannot be seamlessly applied in different countries. Instead, more robust criteria must be established to designate flood or other risk zones.

Land value

The designation of the risk zones has economic implications for the local community. If insurance rates were actuarially correct, the annual payment for risk insurance on a specific hazard should equal the expected annual damage from that hazard. For USNFIP, Chivers and Flores (2002) estimated that the present value of risk insurance premiums ranges from 2 percent to 19 percent of the value of the covered structure and contents. In a perfect market with perfect information, the expected present value of a hazard's damages would be capitalized into the market value of the property located in the risk zone of that hazard. A number of studies have found that this is sometimes true, but often is not true (Chao *et al.* 1998).

Past empirical research has come to varying conclusions about how hazardous risk zoning and development regulations affect property values and development potential. Several studies analyzed by Montz and Gruntfest (1986), Tobin and Montz (1988) and Evatt (2000) come up with conflicting evidence as to how flood zones affect property values. On the one hand, Barnard (1978), Donnelly (1989), Holway and Burby (1990, 1993) argue that floodplain regulations lower property values for undeveloped floodplain land. Prices are lowered because building costs in the floodplain are higher due to regulations that require the elevation of new construction above the level of a 100-year flood. Recently, Troy and Romm (2004) found that the insurance has an effect in decreasing the land value because premiums and the cost of flood insurance are calculated negatively in purchase prices and price differences also include those non-insurable costs. Similarly, Harrison *et al.* (2001), Dehring (2006a) and Bin *et al.* (2008) found a similar effect: the flood insurance program decreased property values.

Other studies have found that flood hazards have no effect on land values. Two early studies that specifically examined the effects of floodplain regulations, as well as flood hazards, on land values have concluded that the regulations have no effect on the value of developed property in floodplains (Damianos and Shabman 1976; Muckleston 1983). Recently, Morgan (2007) found that subsidized flood insurance helps prop up demand and housing prices in floodplain areas are higher than in non-floodplain areas. In general, all natural hazard insurance schemes have an element, more or less large, of cross-subsidy (e.g., people on the flat are subsidizing landslide damage insurance for people on slopes; and people on slopes are subsidizing flood insurance for people on the flat).

Pompe and Rinehart (2008a) argue that the USNFIP provides subsidies to some homeowners in flood prone areas. Without such insurance subsidies, it would be too expensive for many people who already live there to continue to live there. Demand for subsidized homes is higher than it would be in the absence of such subsidies. Power and Shows (1979) made a similar argument that the USNFIP increases the demand for land that will eventually receive subsidized flood insurance premiums, once the floodplain management maps are completed for that land. Shilling *et al.* (1989) proposed that the USNFIP increases the land value for those homeowners with subsidized premiums. Any new home construction after flood-plain management plans are implemented is subjected to actuarial rates instead of the subsidized rates imposed on already existing homes. Hedonic regression techniques demonstrated that subsidized homeowners pay 4 percent less than homeowners with actuarial rates, resulting in a $4 billion wealth transfer nationwide. Zahran *et al.* (2009) argued that insured areas are enticing to homeowners because of the amenities they offer and they are deterred by the hazard of a flood "only to the extent that they perceive a significant flood risk." Similarly, Helvarg (2005) criticizes the sense of security that flood insurance creates that allows mortgages for high risk properties in flood zones. Bagstad *et al.* (2007) contend that flood prone areas that are undeveloped sell at lower costs, but once developed the value is increased. Shrubsole and Scherer (1996) focused on flood policies in Canada and found no conclusive evidence for how land value was affected by floodplain regulations.

The empirical studies, in general, suffer from several key weaknesses, which probably explain the divergent conclusions. First, the studies explicitly controlled for only a few of the factors affecting land value. Second, the expected effects of flood hazard and floodplain regulations on the development potential of land and construction cost should be capitalized into the value of vacant land. Many of these studies examined developed residential properties, not vacant land.

Extending the analogy from floods to other natural hazards, neither the previous research efforts on USNFIP evaluation nor empirical studies for other hazardous program evaluation have specifically examined the cumulative and differentiated effects of multiple hazards on the land values and the likelihood of development, controlling for all other potential effects. Zia and Glantz (2012) thus recommend that this kind of research be assigned priority at the international scale to inform risk insurance-based policy designs and to facilitate the context-sensitive designation of risk zones.

Damage reduction

It is not certain whether risk zoning-based building regulations have reduced damages. For the example of USNFIP, Holway and Burby (1993) provide evidence that USNFIP-mandated building regulations have reduced losses and development in riverine flood hazard areas. Helvarg (2005) suggests that flood insurance slightly decreases flood damages by establishing some minimal standards, which are otherwise considered inadequate in the case of coastal flooding. Luechinger and Raschky (2009) argue that government mandated insurance appears to compensate for any loss sustained in the housing market price of the property. Botzen et al. (2009) propose that flood insurance (not specifically mandated) would be effective at reducing flood damages because survey results show that property owners would respond to incentives to improve flood mitigation construction in exchange for reduced premiums or increased coverage.

On the other hand, Montz and Gruntfest (1986), Shilling et al. (1989), Bagstad et al. (2007), and Pompe and Rinehart (2008a, b) argue that the USNFIP does not reduce flood damage because it increases demand for homes in flood-prone areas, where premiums collected are insufficient to compensate for damage payments. Furthermore, Richman (1993) posited that reductions in infrastructure costs as well as affordable insurance premiums through the USNFIP cause more people to move to flood prone areas, which creates more vulnerability to damages, especially when people do not improve the structure of their homes because the government continues to offer insurance and disaster relief even when homes are damaged. Carolan (2007) argues that the program is inefficient because flood damage is occurring in areas outside the 100-year flood zones. Raschky and Weck-Hannemann (2007) argue that USNFIP does not reduce damages because fewer people opt to get insurance against flooding under the expectation that the government will provide financial aid to those who are uninsured or underinsured in the event of a catastrophe. Sarewitz *et al.* (2003) argue that vulnerability reduction policies do not only consider the risk probabilities of events but also focus on how best to reduce potential damage regardless of risk. Understanding the risk alone won't make structures in flood zones less vulnerable, if they are not altered to sustain the floods. The USNFIP needs to utilize both concepts to help move construction out of areas with high risks of flooding and to encourage better building codes to reduce vulnerability. In summary, past empirical research shows conflicting evidence whether risk zoning-based building regulations increase or decrease damages. Zia and Glantz (2012) recommend that if risk zones do not reduce damages, an investigation must be carried to identify the causes and proper modifications must be made in government policies and strategies to stimulate the development of more resilient buildings.

Land-use planning

The majority of the workshop participants and expert interviewees agreed that if the communities wish to further reduce the rate of increase in the occupancy of hazardous risk zones, they must supplement the building construction requirements (such as building elevation requirements proposed in the USNFIP) with land use regulations limiting new development. The federal governments (and, for that matter, local governments as well) have however neglected to use local land use controls to keep people and buildings away from hazards and, thus, hold down hazard damages.

Traditional land use policies are focused on providing infrastructure and meeting sewer and water needs without explicitly treating hazard mitigation in land use zoning (Tobin 1999). In contrast, Blanchard-Boehm *et al.* (2001) argue that flood mitigation and flood risk reduction have been primary drivers of land-use planning. Recent trends in neo-urbanist movement suggest that smart growth and integrated development plans, which are primarily aimed at growth management and urban sprawl reduction, explicitly consider hazard mitigation aspects in designing resilient communities (Stevens *et al.* 2010).

The broader land use controls have been planned in South Asian countries but they have not been systematically implemented due to a host of political, economic and development problems. Even if some local, provincial, or state governments have required risk zones to be left alone (i.e., no human settlements), landless farming communities and urban poor migrate to those risky areas and establish so-called illegal land developments (Hameed 2005). Local and state governments in South Asia do not have enough resources to stop these land use developments, even if laws or regulations exist on paper. Implementation failures of such land use policy regulations in developing countries are complicated by the problem of corruption (Zia 1999).

In summary, land-use plans are neither implemented as planned nor do they typically include hazard mitigation planning as explicit criteria for designing land-use zoning regulations. Zia and Glantz (2012) support the broader use of land-use planning for long-term mitigation and regulation of hazards. Adaptive policy mechanisms that respond to ground realities and forecast uncertainties could facilitate such changes in land-use planning.

Forecast uncertainty

The USNFIP provides an interesting example of policy implementation for managing flood risk by using flood frequency estimates to designate flood risk zones. Olsen (2006) argues that the assumption behind traditional flood risk analysis is that climate is stationary, but anthropogenic climate change and better knowledge of inter-decadal climate variability challenge the validity of the assumption. Olsen (2006) reviews several alternative statistical models for flood risk estimation that do not assume stationary climate. Although currently out of favor, Olsen (2006) argues, hydro-meteorological models have been used for engineering design as alternatives to statistical models and could be adapted to different climate conditions. Hydro-meteorological models are thus proposed as scientific and objective ways to designate risk zones in flood prone areas.

Pielke Jr. and Downton (2000) correlated U.S. flood damage data (1932–1997) with the precipitation data and found that precipitation measures significantly explain variation in flood damages. The growth in recent decades in total damage is related to both societal factors and climate factors. These findings suggest that climatic changes causing precipitation variability shifts will more likely increase societal damages outside the risk zones established through data from the past 100 years unless anticipatory adaptations through appropriate policy mechanisms are introduced.

Another problem with risk zoning is that climate variability causes uncertainty in the hazard's frequency estimates, such as flood frequency estimates used by USNFIP regulators. Greater uncertainty in actuarial rates could cause a private insurance company to restrict coverage or raise premiums to account for potentially greater risk. However, local or national governments are different from private insurance companies. Raising rates by governmental agencies, such as the U.S. proposals to improve the USNFIP's financial health, could have an adverse effect on other federal disaster relief costs, such as Small Business Administration loans or FEMA disaster assistance grants (Government Accountability Office 2001) and cause some policyholders to cancel their coverage.

Higher premiums thus do not seem to be a viable option to account for the additional uncertainty in hazardous risk estimates resulting from climate variability. This implies that climate change risk in the risk insurance programs will have to be borne by taxpayers. In many societies, taxpayers are already burdened with over-taxed fiscal policies, and adding hazardous taxes will be opposed for political reasons. Zia and Glantz (2012) argue that there is no single best method to re-distribute the risk caused by forecast uncertainty. Rather, iterative deliberative mechanisms must be institutionalized to decide who (governments, insurance industry, or citizens) will bear the increased risk premiums necessitated by internalizing forecast uncertainty.

Costs of accurate maps

Meenar *et al.* (2006) present the problem of "accurate" zoning in the context of disaster management and land use planning issues. They argue (2006: 31) that, to manage flooding and flood insurance policies, communities must be able to measure floodplains correctly. About 20,000 communities have used FEMA floodplain maps for the past 30 years. But many existing maps are out of date by decades and do not reflect today's actual floodplain boundaries. Up-to-date maps are needed to ensure that flood insurance programs are more closely aligned with actuarial risk, encourage wise floodplain management, and increase the public's flood hazard awareness.

Earlier, Power and Shows (1979) found that many reports have been received from communities that the maps are inaccurate and take too long to create. However, no claims have won in court. Arnell (1984) discusses two methods: a "detailed" method that is more accurate but costlier, and an "approximate" method that is less accurate but also less costly. Parker (1995) suggests that local governing authorities rely on floodplain maps which are often inaccurate and unreliable. They have difficulty in restricting the development of floodplains. Burby (2001) notes that the flood zones specifically exclude areas that experience problems due to ineffective storm drainage. Omitting areas that experience flooding regularly maps the zone inaccurately. Carolan (2007) finds that the flood plain maps do not always take into account the changes caused by new development, causing maps to be inaccurate. This is complicated by the fact that about 60 percent of maps are at least ten years old.

Many developing countries cannot afford high-resolution satellite maps, like the ones used in the U.S., for floodplain management. As a cheaper substitute, Sanyal and Lu (2006) created a hazard index on a map of less resolution. Variables such as flood frequency, population density, transportation networks, access to potable water, and availability of higher ground, were mapped in this hazard index map. Overall, generation of accurate risk zone maps is not a one-shot task. Rather, due to the dynamic nature of socio-ecological systems, even 100-year floodplain maps need to be dynamically updated, which raises the costs of these programs. These costs are prohibitively high for developing countries, but worthy of international development investment for long-term capacity building in the developing countries.

Modifiable Area Unit Problem (MAUP)

The Modifiable Area Unit Problem (MAUP) has been recognized as a major problem in risk management and policy science literature that also affects the estimation methods of risk zones. Risk assessment models normally require the integration of pixel-based environmental hazard data with area-based socio-economic data, which poses problems of MAUP, also known as ecological fallacy (Openshaw 1984). MAUP arises due to the scale effect and zoning effect when areal units are aggregated to form units of different spatial arrangements.

Fotheringham and Wong (1991) provided strong empirical evidence on the unreliability of multivariate analysis undertaken with areal socio-economic data at different zone levels or spatial scales. Similarly, the estimation of societal vulnerability from hazardous risk will require the integration of socio-economic data with hazardous risk data. The choice of scale by policy implementation agencies will affect the determination of risk zone boundaries, which can pose inconsistency problems for rank-ordering risk zones based on their societal vulnerability and/or determining appropriate insurance premiums based on societal income groups.

One of the possible solutions to deal with MAUP is to gather data at very fine resolution or at highly disaggregated scale (Chen *et al.* 2003), but this is a very costly proposal for developing countries. Solutions have also been offered to analyze data at multiple scales, such as hierarchical models (Hansen and Bausch 2005), or cross-scalar models (Adger *et al.* 2005b). Most of these possible solutions of MAUP require gathering individual/household level socio-economic data, which is often difficult due to privacy and cost-related issues. In summary, we contend that MAUP makes it impossible to determine risk zones based on any single best technical solution. Risk and disaster management experts must recognize this limitation.

Winners and losers

The choice of different zoning criteria results in the determination of risk zones that can potentially benefit some stakeholders and harm others. Participants in the Islamabad workshop presented strong arguments about winner–loser effects of risk zones. Policy decisions about the criteria of risk zones, as well as land-use zones, directly affect the distribution of wealth. Since there is no single best technical method (as shown in the MAUP problem above), the choices on risk-zoning criteria are social and political in nature. There are contested arguments in the literature about who wins and who loses by establishing USNFIP type of programs.

Power and Shows (1979, 1981), Shilling *et al.* (1989), Daniel (2001), Pompe and Rinehart (2008a, b), Klein and Wang (2009), and Levy (2007) argue that the winners are the people receiving government subsidized premiums to live in known highly flood prone areas, while taxpayers are losers. According to the charity hazard theory (Raschky and Weck-Hannemann, 2007), the winners are the uninsured and underinsured who pay no flood insurance premiums, but receive financial aid in the event of a flood. The losers are the taxpayers who bear this financial burden and the flood insurance policy holders who pay higher premiums than they would if these uninsured were in the pool.

Holway and Burby (1990, 1993) argue that the municipalities seem to be losers in that land values in floodplains are decreased and in turn the tax base is reduced. Harrison *et al.* (2001) argued that winners are municipalities who appear to be over-valuing properties in SFHA and, therefore, are collecting more taxes than is reasonable. Griffith (1994), Burby (2006) and Carolan (2007) suggest that local governments are winners so long as the federal government assumes the majority of the cost in building enforcement, flood mitigation and damage payments.

In the case of India, Mohapatra and Singh (2003) argued that flood insurance is only popular in the urban areas, so residents in the urban areas benefit substantially from the insurance coverage while rural populations lose in the long run. Olsen (2006) argued that the taxpayers will absorb any risk due to climate change uncertainty, so in the long run, winners are the flood insurance customers while the taxpayers are the losers. Gopalakrishnan and Okada (2007) argue that the society as a whole is a loser due to systematic failures in disaster management. In particular, losers are those people who are poor and live in low-lying areas throughout the world. Cummins (2006) suggests that government mandated risk insurance programs such as USNFIP lock out insurance companies from the market; hence insurance companies are the losers. Chivers and Flores (2002) argue that home buyers in flood zones are losers as they typically do not know about USNFIP requirements until closing and are unable to factor USNFIP requirements into the offer. Similarly, Evatt (2000) suggests that developers and local governments with increased property tax base are winners of public insurance schemes. Bell and Tobin (2007) suggest that winners are those who respond positively to the flood risk information being conveyed and the losers are those who do not respond to flood insurance risk information. In summary, whether government-subsidized or micro-insurance programs, each method or criterion to determine risk zones results in a different distribution of winners and losers. Re-distributional consequences of risk governance mechanisms must be made transparent, while acknowledging that such information is contested.

Single versus multiple hazards

Another major challenge identified with risk zoning is the issue of overlapping risk zones in the case of multiple natural hazards in a given area. Many workshop participants and interviewees argued that instead of developing single-hazard risk zones, it would be more appropriate to develop multi-hazard risk zones. However, multi-hazard risk zones lead to the problem of integrating damages from multiple hazards over heterogeneous time-scales for determining actuarial rates. For example, flood insurance in Spain and France falls under a catastrophe coverage program required for all property owners. The French and Spanish model has been criticized for inducing moral hazards through its lack of incentives in stimulating robust building designs (Linnerooth-Bayer *et al.* 2003; Klein and Wang 2009).

Friedman *et al.* (2002) and Keeler *et al.* (2003) propose that the erosion insurance could be combined with the flood insurance. Another insurance program that is offered with flood insurance is wind damage through private insurers (Kunreuther 2008). In India, private insurance companies are combining flood insurance with other risks (Mohapatra and Singh 2003). Arnell *et al.* (1984) report that flood insurance is generally a part of regular insurance policies in

Britain along with storm, fire and theft insurance. Pompe and Rinehart (2008b) report that some southern states like Florida offer state subsidized hail and wind damage insurance because the USNFIP does not. In summary, Zia and Glantz (2012) argue that the establishment of multi-hazard zones is desirable from a policy standpoint over single-hazard risk zones. While the French and Spanish model has incentive and moral hazard problems, design of multiple-hazard risk zones could be manipulated to minimize moral hazards.

Cross-jurisdictional administrative boundaries

Blanchard-Boehm *et al.* (2001: 26) note that flood insurance zones spill over traditional boundaries including historical water rights governed by "prior appropriation doctrine." Burby (2001) suggests that zones do spill over, though the U.S. program allows for local municipalities to go above and beyond the USNFIP requirements. Bagstad *et al.* (2007) note that historically, in the United States, the Army Corp of Engineers (USACE) has managed coastal and other major waterways that cross assorted boundaries. Currently, USACE is in the process of transferring control to local municipalities. Carolan (2007) sums up the cross-boundary problem in terms of enforcement issues. Though the flood zones and the regulations surrounding them are created by the federal government, enforcement is left to local governments and often doesn't happen. Many workshop participants and interviewees recommended that inter-organizational governance networks for risk zoning should be established and periodically revisited. Both expert and citizen groups should be represented in these inter-organizational governance networks, as well as governmental agencies and non-governmental agencies should be given adequate representation. The problems of assigning accountability in these inter-organizational governance networks remain unresolved (Koliba *et al.* 2010) and need to be studied further in the context of enforcing programmatic regulations.

The politics of scale, whether temporal or spatial, is observed, as argued in Chapter 2, for the development of climate change mitigation and adaptation policies. In particular, Chapter 2 presented examples of politics of scale in mitigation or adaptation policies. Next, in Chapter 3, I venture to discuss politics of scale for two specific policy mechanisms that are expected to be carried over from the Kyoto framework to a post-Kyoto framework: The first one is CDM and the second one is REDD+.

3 The politics of scale II

Synergies and trade-offs in complex systems

Building on the previous chapter, this chapter advances the application of the politics of scale lens to assess synergies and trade-offs caused by introducing various policy interventions in complex systems. Conventional climate policy tools are demonstrated to be inadequate for assessing synergies and trade-offs in complex systems, such as the inconsistencies and abstractions that are generated by the conventional notion of measuring opportunity costs for assessing trade-offs in the coupled social-ecological systems that effect interactions between atmosphere, ocean and terrestrial land. First, a generalizable adaptive management approach is presented to assess synergies and trade-offs in the context of biological conservation, with the note that tropical deforestation is causing 17 to 25 percent of global GHG emissions. In the following two sections I apply this generalizable approach to demonstrate the politics of scale in designing and governing REDD+ and CDM policy mechanisms, respectively. I focus on REDD+ and CDM, as both of these policy mechanisms are expected to be part of a post-Kyoto climate governance regime.

Assessment of synergies and trade-offs in complex systems

Before I describe a formal method to assess synergies and trade-offs of various policy and governance interventions in complex systems, I'd like to briefly describe the history of biological conservation policy vis-à-vis economic development. In particular, I use this example to describe the level of complexity that policy-makers are dealing with when introducing climate change mitigation policies such as CDM and REDD+, which are also touted as "synergistic" with climate change adaptation, air quality management, poverty reduction and biodiversity protection policy goals. As argued earlier, the WCED (1987) definition of sustainable development assumed a win–win relationship between conservation and development goals, an assumption that has not withstood the empirical test in the last twenty-five years of unfettered development and continuous loss of tropical forests, which are contributing anywhere from 17 to 25 percent of GHG emissions per year. In fact, the history of conservation policy interventions in tropical countries, and elsewhere, reveals a complex patchwork of synergies and trade-offs between conservation and development goals. Next, I briefly present this history and then present a generalizable adaptive management approach to assess and navigate synergies and trade-offs of various policy interventions in complex systems. This adaptive management approach is then applied to assess the synergies of REDD+ and CDM vis-à-vis other non-carbon management policy goals that are typically touted as "win win win" or "synergistic" effects of pursuing these climate policy mechanisms.

Contemporary discourses on conservation and development identify three over-arching paradigms that have been historically deployed to design policy interventions for sustainable protection of global biodiversity, including tropical forests (Blaikie and Jeanrenaud 1997; Brown 2002):

1   the classic approach;
2   the populist approach; and
3   the Neo-liberal approach.

The classic approach framed the decision problem by treating local community welfare and development as directly opposed to the objectives and practices of biodiversity conservation. Development was perceived as the key detrimental factor for protecting biodiversity. The classic approach led to the policy prescription of "fortress conservation," implying traditional top-down exclusionary approaches to protected areas (Anderson and Grove 1990). The classic approach failed, however, in protecting the biodiversity, as has been recorded by the overwhelming evidence of global biodiversity loss during the twentieth century (Wells 1992; Wilson 1989, 2002). Essentially, the classic approach failed because it alienated local communities (Schroeder 1999) and it was perceived as a drain on the scarce resources of many developing countries that happen to contain the major biodiversity hotspots (Pimm *et al.* 1995, 2001).

In response to the failure of classic paradigm, populist and neo-liberal paradigms have developed concurrently during the last two decades. Both populist and neo-liberal paradigms stress complementarities and trade-offs and reject the notion of strict conflict between conservation and development. The populist paradigm treats participation and empowerment of local communities as a key mechanism to protect biodiversity on a sustainable basis (Hulme and Murphree 1999). The populist paradigm argues for the development of buffer zones, or biosphere reserves, for creating employment and development opportunities for local communities. There are however problems with this paradigm: (a) biodiversity loss may be accelerated by this kind of policy prescription (e.g., see Langholz 1999); (b) public participation may be used as an alibi to further strengthen the power of timber mafias and government agencies (e.g., see Agrawal and Gibson 1999; Cleaver 1999).

The neo-liberal paradigm treats institutional, market and policy failures as the key causes for the loss of biodiversity. The neo-liberals promote the policy prescription of adding economic value to biodiversity and correcting institutional and market mechanisms through policy driven shifts of property rights of the biodiversity hotspots (Adger *et al.* 2001). Essentially, neo-liberals promote the "joint" or "communal" management (or co-management) of biodiversity resources requiring protection. The key problem with this paradigm is, however, the need for a major shift in the institutional structure of developing countries, which in turn opens up another Pandora's Box: how to re-allocate property rights? Further, how to place economic value on biodiversity, given the known problems of positive discount rates, which asymptotically diminish the net present value to zero of any long-term project with a lifetime of 100+ years, as discussed in Chapter 2 with respect to the politics of temporal scale. Finally, who will ensure that communal management will not fall prey to the power structures that caused the failure of the populist/communitarian paradigm?

Figures 3.1, 3.2, and 3.3 show the "problem maps," respectively, for each of the three paradigms: classic, populist and neo-liberal. These problem maps demonstrate the differences in the "frames" of the respective decision/policy problems through the representation of stocks or state variables (rectangles), flows (ovals) and the direction and polarity of hypothesized effects (arrows and signs), their respective policy solutions/prescriptions (diamonds) and operational features of management/policy implementation (hexagons).

Having described the protracted history of failure in biological conservation policies, a generalizable approach to assess synergies and trade-offs is described. This generalizable approach starts with the humble acknowledgment that there is no single correct way to frame a policy problem, as has been demonstrated by Rittel and Webber (1973) and extensively discussed in the context of environmental management (Norton 2005). Even the deployment of so-called multiple criteria decision-making (MCDM) models, which allow the assessment of synergies and trade-offs for various policy and governance interventions, requires "taming" the wicked policy problems by framing the problem to be evaluated.

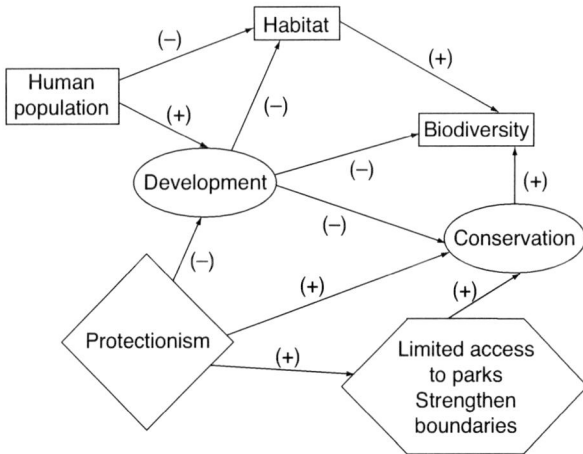

Figure 3.1 An illustrative problem map of the "classic paradigm" for conservation policy.

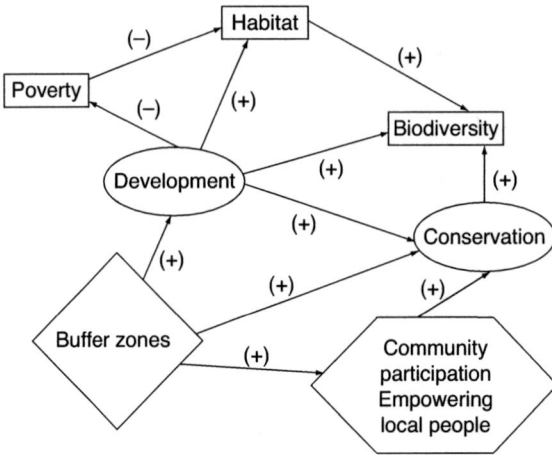

Figure 3.2 An illustrative problem map of the "populist paradigm" for conservation policy.

The MCDMs have been considered for explicit consideration of assessing synergies and trade-offs in the context of biodiversity protection and development related issues. Guikema and Milke (1999), for example, provide a broad overview of decision models used by different conservation agencies (including United States National Park Service, IUCN) and conclude that the value trade-offs is not adequately treated by these agency models. They propose a utility-based MCDM, but they frame the policy problem from the perspective of agency decision to allo-

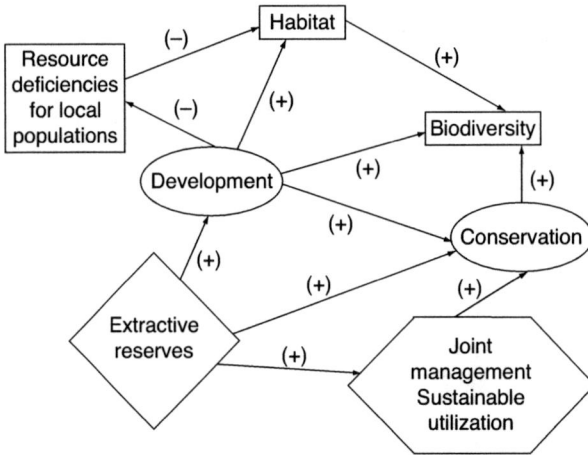

Figure 3.3 An illustrative problem map of the "neoliberal paradigm" for conservation policy.

cate scarce resources among alternative conservation projects (also see Faith and Walker 1996 for similar treatment of trade-offs between biodiversity protection and costs). Alternative projects are evaluated on multiple criteria/values, and synergies and trade-offs among the values are modelled through assigning the constant-sum weights to multiple criteria/values. While their proposed methodology is appealing, and probably a step forward in the explicit consideration of value trade-offs and synergies for conservation agency's budgetary decision-making, it does not take into account the frame of local communities other than asking them for their subjective weights on values. Further, their methodology does not treat the problem of meta-weights, i.e., whether each stakeholder gets equal weight in deciding the final weights on values, or some stakeholders get more weight than others depending upon their decision-making power and impact (e.g., park managers versus tourist guides versus poor communities). Meta-decision choices (discussed below) are taken for granted.

Careful consideration of meta-decision choices in MCDM applications are needed to assess synergies and trade-offs in complex systems. Formally, let $A \neq \emptyset$ be defined as a non-empty set or vector of alternatives (also called policies, actions, strategies or feasible solutions) of a decision problem. The set of alternatives is always non-empty because the alternative of "no action" is always an alternative in any decision problem. Further, in a most generalized sense, let a multi-criteria outcome function f be defined as follows:

$$f: A \rightarrow R^x \tag{3.1}$$

Each function $f_k: A \rightarrow R$ with $f_k(a) = z_k (k \in [1, \ldots, x], a \in A)$ and $f(a) = (z_1, \ldots, z_x)$ is defined as a multiple value function. In the most general sense, $\varphi = (A, f)$ is defined as a multiple criteria decision-making (MCDM) problem, wherein φ is a matrix showing a generalized decision problem involving a set or vector of Alternatives A faced by n decision-makers and containing f outcomes. The decision-makers measure the outcomes by z_x values. More specific formulation of decision problems is undertaken by adding future events (e.g., decisions under uncertainty) and/or replacing multi-criteria values with an expected utility function (e.g., economics literature), and so on.

Next, for defining and solving the decision problems in public decision-making contexts, the formal version of four meta-decision problems—choosing the alternative set, criteria set, weighting methodology and method set—are presented in the decision problem formulation context of equation 3.1. A brief review of the meta-MCDM problem is also undertaken to demonstrate that algorithmic solutions to resolve meta-MCDM problems face severe constraints in the case of designing our environments.

Choice of space-time boundaries (and problem bounding)

The first meta-decision problem concerns whether the set of alternative paths A is a finite set (as defined by many expected utility and multiple attribute decision-making (MADM) theorists) or is it infinite (as defined by multiple objective decision-making (MODM) theorists) or is it fuzzy (as defined by fuzzy set theorists). Further, what meta-criteria, such as space-time boundaries of a decision problem, should be used to include or exclude an alternative path from A? What is the logic of establishing space-time boundaries by which an alternative is included in the set of policy and planning alternatives? These questions can be referred to as the meta-decision problem of the alternative set. An example in the context of development-biodiversity trade-offs is the problem of choosing appropriate spatial scale (e.g., eco-system based boundaries or politico-administrative boundaries such as nation states) and temporal horizon (e.g., ten years, 100 years, 1000 years and so on).

Value ambiguity

Which values are/should be included in the criteria set of evaluation to measure the outcomes of our actions? Is the value set compact and closed or is it non-compact and open? Do human societies only care for the values of cost-effectiveness, fairness, efficiency, social justice and environmental preservation in evaluation of any environmental policy decision; or are there/should there be some additional values such as eco-system health/services, animal welfare that are/should also be included in the evaluation process? Concisely, what is the logic of a meta-choice that a value is/should be included in the criteria set of evaluation? In the rest of the book, value ambiguity is referred to as a meta-decision problem of the criteria set.

Formally, the second meta-decision problem concerns the decision as to which value/criteria $z_k(k \in [1, \ldots, x])$ shall be included in the multiple value function of equation 1. Restricting the value set to 1 element concatenates the MCDM problem to a scalar problem. For example, the cost-benefit function concatenates any decision problem with multiple valued outcomes to a single valued outcome. All the values are thus represented by monetary units, which are *commensurable* scalar quantities. In the case of $x \geq 2$, we have a multiple-value decision problem. The meta-decision problem remains: which values shall be included in the evaluation function f to determine the desirability of actions faced by decision-makers?

Choice of weighting functions

How shall the weights be assigned to the pluralistic values and alternative mixes on the basis of which we measure the outcomes of our actions for judging good actions/decisions? This is called the meta-decision problem of weighting methodology.

Formally, the third meta-decision problem concerns which weighting methodology shall be used to weigh the $z_x \geq 2$ values or $a_k \geq 2$ alternates. Should the value (or alternate) trade-offs be set up as a zero-sum game with $\Sigma_{h=1}^{x} w_h \cdot z_h = 1$ or a positive-sum game with $\Sigma_{h=1}^{x} w_h \cdot z_h > 1$? Furthermore, which methodology should be used to ascertain the values of the weights w_h for the criteria z_h (where $h=1,\ldots, x$) or alternate mixes $(a_k!)$? The choice of weighting methodology and assignment of actual weights explicate underlying assumptions about value trade-offs.

Choice of decision rules

Given the multiplicity of decision models and algorithms, policy-makers and planners are confronted with the problem of how to choose which descriptive or normative decision rule/algorithm to apply in a given situation. This is called a meta-decision problem for determining the decision rule set.

Formally, the fourth meta-decision problem concerns which decision rule τ (decision algorithm, decision method) shall be used to solve the decision problem $\varphi=(A, f)$. There is no meta-algorithm that tells the users when to apply one decision algorithm and when the other. Should a cost-benefit analysis, a concatenated version of simple additive weighting (SAW) and expected utility decision rules, be used in choosing climate change mitigation policies that also preserve the current level of biodiversity in the year 2100? Or if a cost-benefit analysis is not able to measure the preferences of the yet unborn future generations, and/or not able to set appropriate future discount rates, should another decision rule, such as back-casting or analytical hierarchy process (AHP) be used to choose the desirable level of GHG emissions that also preserves the value of biodiversity protection?

The choice of a decision algorithm may affect the choice of weights on values and/or alternatives, such as SAW only adds, while the weighted product only multiplies the expected values. There is, however, not a single meta-decision algorithm that lets the policy/decision-makers choose the appropriate decision model for specific environmental design problems.

Norton (2005) has argued that we need adaptive management to solve wicked environmental policy and governance problems. In adaptive management, the environmental policy decisions are modelled at two interactive levels: At the level of action, descriptive analysis is undertaken to ascertain the current state of the world, such as existing environments, institutions and policies, and the outcomes ensuing from current policies/decisions. At the meta-level of reflection, normative analysis, such as the proposed MDMs, is employed to determine the socially desirable values by which outcomes of (current and future) policy actions are measured.

At the meta-level of reflection, the planners and policy designers compare the outcomes measured at the level of action with the outcomes that are deemed normatively desirable within the space-time horizon of environmental policy decisions. At this level, meta-decision choice problems are resolved through iterative experimentation and active collaboration/participation between expert and lay decision-makers. The normative analysis at the meta-level of reflection results in policy prescriptions/recommendations that aim at getting "there" from "here" given all the uncertainty, ignorance and incomplete information. The most important aspect of adaptive management concerns the explicit representation of meta-decision choices on A, x, w and τ. These choices are not made for all the temporal periods that exist between now and the planning horizon; rather the meta-decision choices are revisited and revised at each iterative evaluation of the then current state of environment and policies/institutions.

The adaptive management approaches can be integrated with a range of existing decision evaluation models, environmental impact assessment models, scenario analysis models and so on. Further adaptive management approaches can be used to determine the specific meta-decision choices that were made/assumed while applying these other models.

In an adaptive management approach, participatory decision-making approaches, which are the core of MDMs, are iteratively applied to frame and re-frame policy problems before assessing synergies and trade-offs among different policy interventions. Here, a generalized methodology is expounded. For more details and a field application, please see Zia *et al.* (2011). First, various competing frames of the decision problem $\varphi=(A,f)$ are elicited through explicit stakeholder participation. Second, meta-decision choices from various concerned parties are determined through carefully designed participatory decision seminars (or focus groups or workshops). Third, the elicited meta-decision choices are used to do a descriptive and a normative analysis of the environmental design problem. Fourth, a gap analysis is undertaken to estimate the gaps on various value dimensions of desirable outcomes between current and future environments. An uncertainty analysis and a sensitivity analysis are also undertaken at this stage. Fifth, policy analysts/planners devise specific recommendations for an informed discussion among all the concerned parties. Sixth, this process is iteratively repeated at periodic intervals until a stable set of trade-offs and synergies for various policy interventions start to emerge from multiple iterations.

The aim of the proposed six-step participating MCDM methodology is not to declare that a new meta-algorithm has been created that can be used to permanently resolve the wicked environmental design problems. Rather, the proposed adaptive management approach aims at illuminating the limitations of the current evaluation models as well as providing general stakeholders with specific information on various meta-decision choices, such as space-time boundaries of decision horizons, value trade-offs, emerging/new alternatives and newer decision models emerging over multiple iterations spanning multiple generations. More importantly, such participatory methods could illuminate context specific/scale specific synergies and trade-offs among various policies, including but not limited to mitigation and adaptation policies.

REDD+: synergies and trade-offs between climate change mitigation, biodiversity conservation, and food security

The recent evolution of international climate policy mechanism that was initially known as "Reduced Emissions from Deforestation" (RED) to "Reduced Emissions from Deforestation and Forest Degradation" (REDD) and its variant

REDD+ provides an example of the limits of predictability in predefining the phase space of managing tropical social ecological systems in a short time-span of 16 years since international climate policy negotiations began under the auspices of UNFCCC. I use the evolution of REDD+ policy mechanism as an example to shed light on how and why global governance on climate change, food security and biodiversity loss has "synergistically" converged around the international policy goal of conserving tropical forests. Further, complementing REDD+, recently the United Nations Environment Programme (UNEP) in CBD (Convention for Biodiversity) COP 10 has promoted TEEB (The Economics of Ecosystems and Biodiversity) approach as another policy mechanism to reduce deforestation in tropical countries. Both the REDD+ and the TEEB approaches follow the underlying "command and control" model of predefining phase spaces of social ecological systems and then imposing a "market" model to incentivize tropical forest conservation. Unfortunately, these models ignore the implications of international free trade, mandated under WTO policy mechanisms. Further, social and ecological values that could not be priced in dollar/monetary terms are also not represented in such models.

In general, REDD+ and TEEB have been conceptualized as a "win win win" policy mechanism for mitigating climate, protecting biodiversity and conserving indigenous culture by institutionalizing payments on carbon sequestration and biodiversity conservation values of ecosystems services from global to local communities. The Union of Concerned Scientists (UCS), for example, asserts that REDD is an option that "not only averts global warming's worst consequences but also generates enormous co-benefits for biodiversity conservation and sustainable development" (Boucher 2008). UN REDD Program (2009a, b) states that REDD policy is an effort to create a financial value for the carbon stored in forests, offering incentives for developing countries to reduce emissions from forested lands and invest in low-carbon paths to sustainable development. The UN-REDD Program focuses on the multiple benefits that can be provided by REDD, specifically the ecosystem benefits. According to UN-REDD Program definition, the multiple benefits of REDD, in addition to its contribution to climate mitigation, include forest conservation, which will protect biodiversity and provide ecosystem services.

The position of some other stakeholders is a bit more nuanced than the UN's REDD program. For example, according to the Climate Action Network (2007), emissions from tropical deforestation must be reduced to keep global temperatures from increasing more than 2°C by 2050. Global warming will contribute to the destruction of tropical forests overtime and will negatively impact forest biodiversity. REDD+ projects can help preserve biodiversity which improves the resiliency of the forests in turn. If an international agreement includes provisions for REDD+, the systems must be developed at the national level to address the issues of leakage and calculating baselines. Industrialized countries would compensate the tropical countries for their opportunity costs and provide funding for capacity building (Climate Action Network 2007). In particular, governments need help monitoring and measuring degradation activities because it requires different technology and is more expensive than measuring deforestation.

Although the idea behind REDD+ and TEEB policy mechanisms may be simple (compensate developing countries for sustaining tropical forests), actors in the policy arena do not necessarily agree on how REDD+ and TEEB policies should be designed, the costs and benefits of implementation, and whether REDD+ and TEEB are the best approaches to reduce emissions. Tropical deforestation is an inherently complex problem, and many efforts and policy approaches, as outlined on page 50, have failed so far. Any attempt to solve the deforestation problem generates unintended consequences that will undoubtedly impact the implementation of solutions in the future (Hirsch *et al.* 2011). Global deforestation and degradation is the result of an increasingly global economy; the demand for particular goods from developed countries has encouraged the conversion of tropical forests to other agricultural uses, including growing soybeans, raising cattle, and producing crops for new biofuel technologies (Mardas *et al.* 2009). Far from being a win-win, REDD+ and TEEB policies will be interventions in a highly complex system, and will inevitably involve trade-offs; therefore it is important to question "win-win-win" discourse.

In general, REDD+ and TEEB payments are calculated by measuring an opportunity cost of foregoing deforestation that generates a variety of ecological valuation puzzles. The opportunity cost of REDD+ and TEEB is defined as the cost that will be incurred in retaining existing tropical forests. "Retention means sacrificing the opportunities that would be gained by converting the forest to other uses, such as crops or pasture" (Boucher 2008). Opportunity cost is thus considered as the minimum amount that would need to be paid to keep the land in forest. A major thrust of economic analysis of REDD+ and TEEB programs has been to calculate this opportunity cost ($), which is typically divided by carbon density of the forest (e.g., tons per hectare) to calculate the minimum cost of REDD+ that is expressed in units of money/area (e.g., $/hectare). Similarly biodiversity values are calculated to derive TEEB payments. The measurement of opportunity costs requires heroic assumptions, such as future prices of soybean or other crops that could have been grown on the deforested lands. It is due to these inherent uncertainties about future counterfactuals (i.e., future phase spaces) that different economic models come up with different opportunity costs for REDD+. Global models produce REDD+ cost curves that are typically higher than regional or local models (Nabuurs *et al.* 2007; Boucher 2008). While this inconsistency is still being debated, this version of opportunity cost ignores other costs associated with implementation of REDD+ and TEEB such as administration or capacity building costs. There have been no reliable global studies that estimate administration or capacity building costs. A more subtle methodological problem in the calculation of REDD+ costs in terms of carbon abatement (e.g., $/tCO$_2$) concerns about the underlying assumptions whether a single buyer or a cartel is assumed to set the market price (typically area under the abatement curve), or a global carbon market price be used to determine the price of REDD+ payments. The uncertainties about the non-linear curvature of carbon abatement curves further complicates the calculation of REDD+ payments. Similar problems exist in estimating biodiversity loss functions for designing TEEB.

Notwithstanding difficult challenges and uncertainties in measuring carbon abatement curves and opportunity costs, the calculation of carbon densities poses even more daunting challenges (Ramankutty *et al.* 2007). It is due to these measurement problems that the overall contribution of REDD+ to global climate change GHG flux is still contested. According to the Intergovernmental Panel on Climate Change (IPCC)'s fourth assessment report (Nabuurs *et al.* 2007), tropical deforestation contributes about "20 percent" of global GHG emissions. There are, however, important methodological issues that underlie this "narrative" of 20 percent estimate. DeFries *et al.* (2002) and Achard *et al.* (2002, 2004) used remotely sensed tropical deforestation data to estimate carbon releases and found that Houghton (2003a, b) and Fearnside (2000) have overestimated carbon emissions from land-cover change by up to a factor of two, mainly because of different estimates of the rates of tropical deforestation. The differences among these studies can be ascribed to modelling different geographic ranges and time periods, different types of land-cover changes, different assumptions about land-cover change and different carbon cycle models/fluxes. There are considerable scientific uncertainties about quantification of several key elements for an accurate and complete analysis of carbon density estimates in tropical forests. These include rates and dynamics of land-cover change, initial stock of carbon in vegetation and soils, mode of clearing and fate of cleared carbon, response of soils following land-cover change, influence of historical land-cover legacies and the representation of processes in the models used to integrate all of these elements (Ramankutty *et al.* 2007). Recently, Van der Werf *et al.* (2009) argued that 20 percent estimate needs to be revised downwards to 12 percent as GHG emissions from fossil fuels are increasing faster than tropical deforestation. Differences in carbon density calculations will inevitably effect the calculation of REDD+ payments, which can be potentially used by the developed countries to underestimate REDD+ payments or by the developing countries with tropical forests to overestimate the REDD+ payments.

Calculating a baseline upon which to compare subsequent activity is essential to valuing reductions in carbon emissions. Depending on the method of calculation used, data may not be available for all countries. Reference levels can either be based on historical or projected rates of deforestation in a particular country or region. Historical baselines aim to measure any reductions in deforestation below past trends. A variation on this approach is the historical adjusted baseline; this method incorporates factors that may impact the rate of deforestation in the future due to development pressures and adjusts the historical baseline accordingly. One caveat of the historical adjusted baseline is that it could actually provide financial incentives to countries that achieve a net increase in deforestation, as long as the rate of deforestation is below the adjusted baseline. Projected baselines require the most sophisticated data because the method relies on econometrics to assess the future rate of deforestation in a country or region based on the social and economic driving forces of deforestation. Again, the projected baseline approach is susceptible to allowing financial rewards to countries that actually contribute to a net gain in deforestation (Parker *et al.* 2009).

An important REDD+ and TEEB policy design problem is whether carbon (biodiversity) offsets or credits are issued at the national or the project level. On the one hand, national-based approaches are favorable because REDD+ (and TEEB) efforts will be weakened unless incentives to convert forests to other land uses are eliminated (Forests Dialogue 2008). A national approach is also more conducive to establishing baselines and addressing leakage; however, project-based and/or community based approaches may be appropriate if a country is not ready to implement a national REDD+ policy and take on the responsibilities of monitoring forest activity (Climate Action Network 2007). Further, it has been empirically demonstrated through an analysis of the International Forestry Resources and Institutions (IFRI) database (available at http://www.sitemaker. umich.edu/ifri/home) that community based governance of forest commons is more effective in the long run than national level governance approaches (Chhatre and Agarwal 2009). On the other hand, a national-based approach to developing and implementing REDD+ and TEEB policies is preferable to minimize the possibility of leakage within a country or across the countries. It is argued that monitoring emissions reductions at a national level discourages leakage, or the displacement of deforestation and degradation activities, to other parts of the same country. But there is a trade-off here as national governments are generally in conflict on tenure right issues with many local and indigenous communities. There are significant disagreements even about the definition of local and indigenous communities in REDD+ and TEEB negotiations. In addition to these definitional issues, there are significant power asymmetries and long-standing conflicts between indigenous communities and national governments, and the implementation of REDD+ and TEEB projects by national governments could further exacerbate these conflicts depending upon who is defined as "indigenous" by the national governments for transferring REDD+ and TEEB benefits.

REDD+ policy will undoubtedly require an exchange of resources among relevant actors, but the allocation of REDD+ policy benefits and burdens among actors is not determined at this time. A more thorough/complete review of the projected costs and benefits identifies some incommensurate puzzles in terms of the impacts international REDD+ and TEEB policy mechanisms may have on individual stakeholder groups, especially three target populations: developed countries, developing countries, and local and indigenous people. Developed countries are by-and-large expected to provide the bulk of the financing for any REDD+ and TEEB mechanisms that are put into place. Financing is the crux of the success for REDD+ and TEEB project proposals; without monetary support, developing countries will not be able to effectively reduce emissions from land use practices. Although the pressure to finance REDD+ and TEEB projects could be seen as a burden on the developed countries, it is also a benefit to these actors because it is arguably less costly to pay for REDD+ and TEEB projects in developing countries rather than invest significant financial capital to reduce fossil fuel driven emissions from energy sectors through improving their own infrastructure and technology. The pressure for developed countries to finance

REDD+ and TEEB has also provided them with extreme bargaining power. Developing countries are dependent upon the developed world to support their actions; until concrete figures of anticipated financial compensation are provided, it is unreasonable for anyone to expect developing countries to implement REDD+ and TEEB projects on a broad scale.

Each developing country that chooses to participate and support REDD+ and TEEB policy mechanisms will need to assess their institutional capacity for implementation and monitoring efforts. Other national land use policies may need to be revisited to ensure that the efforts for REDD+ and TEEB are not undermined and land use tenure rights may need to be better defined. This undeniably places the burden for action on developing countries, although not all countries will be eligible to participate. Depending on a country's historical rate of deforestation and the amount of forest cover they have, only some developing countries may be rewarded. It is arguably unfair to countries that have kept deforestation under control previously to be excluded from any REDD+ or TEEB mechanisms now. This approach could also create perverse incentives for countries with high forest cover and traditionally low rates of deforestation to negatively change their land use practices. "The governance and administration of the REDD mechanism will be critical to ensuring the equitable distribution of benefits among and within countries with tropical forests" (Thies and Czebiniak 2008: 4).

Amidst this power tussle among developed and developing countries, it has been widely recognized that indigenous populations are the most vulnerable. There have been growing calls for increased participation of indigenous groups in REDD+ and TEEB deliberations to provide a "narrative" of legitimacy to the policy design process. The REDD+ Social and Environmental Standards Committee, for example, has consistently advocated for all relevant stakeholder groups to be involved in program design, implementation and evaluation through effective consultation or more active participation. There are particularly concerns in those tropical areas where land use rights are already contentious in the wake of colonial and imperial history. One of the central issues with respecting the involvement and perspectives of the indigenous populations is the fact that in many cases they lack legal land rights to ancestral homes and are socially and economically marginalized in their own societies. A significant proportion of the world's forests are owned by states themselves and most fall under state control in some way, even if only in relation to land-use zoning laws. "Government decisions about land-use zoning and forest management often do not take adequate account of the rights of Indigenous groups living in such areas, particularly when these conflict with perceived national interests or opportunities for financial gain" (Barnsley 2009: 50). Indeed, "self-serving central government-oriented REDD plans are already emerging" (Griffiths 2009: 20). Because indigenous value systems are disregarded in the initial stage of conceptualizing the problems, indigenous peoples are excluded from a position in the process of policy formation and administration. This represents a lapse in respect for human rights as it renders a community powerless to design their own future. Indigenous communities left to the mercy of outside interests is a clear example of how the dominant theoretical framework defines the phase space of a problem and therefore controls the *process* of addressing those "predefined" problems. The resulting mechanisms of policy administration preserve the old, unequal power dynamic.

While the implementation policy for protecting the rights of indigenous communities is still being debated, both REDD+ and TEEB face a fundamental challenge from the international trade regulations mandated under WTO arrangements; and food and agricultural subsidies provided by World Bank, regional development banks, and other international organizations promoting standard "western" style of neoliberal governance policies in tropical forested developing countries under the garb of "food security" and "economic development." The phase spaces for REDD+ and TEEB policy mechanisms are designed on such a counterfactual basis that the calculation of opportunity costs, forest densities and carbon densities will remain uncertain so long as free trade and economic growth policies continue to promote the spread of mono-agriculture in tropical forests. These mono-cropping agricultural practices are either promoting the replacement of old growth forests in tropical countries with food system input crops (e.g., growth of soybean and palm oil plantations in tropical forests during the last fifty years) or biofuel energy inputs (e.g., corn for ethanol). In theory, the replacement of old growth forests with mono-crops will conserve the carbon density of the forests; however, in practice, it could trade-off biodiversity (as animal habitats will be lost due to mono-cropping practices) as well as indigenous communities (as their habitats will be lost), perhaps as unintended consequences. There are thus serious trade-offs among REDD+ and TEEB policy mechanisms that are being designed by different arms of the UN system. The phase spaces that are assumed in the design of REDD+ and TEEB policy mechanisms, vis-à-vis WTO alone, cannot claim to completely represent the phase spaces of dynamic social ecological systems that are evolving in tropical forest countries. For example, the uncertainty regarding the impact of global climate change, e.g., shifting of precipitation patterns over the Amazon and other tropical systems, puts into question the efficacy of REDD+ and TEEB investments in the long run, which instead could be used to reduce GHG emissions in fossil fuel intensive developed countries. There is thus large value ambiguity and system uncertainty about the phase spaces of tropical forest countries that require strict boundary assumptions for designing REDD+ and TEEB policy mechanisms and heroic assumptions about the state variables, e.g., global climate change, governing the evolution of tropical social ecological systems. In a nutshell, REDD+ and TEEB represent a classical example of politics of scale, whereby global scale social organizations are devising competing policies to simultaneously promote and stop deforestation in relatively poorer tropical countries. Whether and how these policies will succeed remains to be seen, given the history of failures and disenchantments in tropical forest countries.

CDM: synergies and trade-offs between clean energy, international development, and climate change mitigation

The Clean Development Mechanism (CDM) is one of the "flexibility" mechanisms defined in the KP. The CDM has two main goals: cost-effective compliance with the Kyoto Protocol for developed (Annex B) countries through greenhouse gas (GHG) emission reductions in developing (non-Annex B) countries, and contributing to sustainable development in non-Annex B countries (Bakker *et al.* 2011). The CDM allows industrialized countries to invest in emission reductions wherever it is cheapest globally. Between 2001 and 2012, the CDM is expected to produce some 1.5 billion tons of carbon dioxide equivalent (CO_2-E) in emission reductions, where most of these reductions are being achieved through renewable energy, energy efficiency, and fuel. While 1.5 billion tons of CO_2-E in emission reductions are significant in and of themselves, they pale in *scale* to the hundreds of billion tons of CO_2-E produced by developed countries, as discussed in Chapter 2.

The goal, and considered success, of CDM is to create a more "cost-effective" method for industrialized (Annex B) countries to comply with the Kyoto Protocol by fulfilling their mandated decrease in GHG emissions through the purchase of carbon credits by implementing carbon decreasing projects in developing countries (Nazifi 2010). The carbon credits exchanged through CDM are referred to as certified emission reductions or CERs. One CER is equal to 1 ton of carbon dioxide (UNFCCC 2011). CDM is cost effective since emissions trading schemes theoretically lead to a decrease in carbon prices. The argument goes that it is relatively less expensive for developing countries to make the adaptations necessary for decreased carbon emission so it is worthwhile for industrialized countries to "sponsor" these CDM projects in exchange for emission reduction credits (Nazifi 2010; Zhang *et al.* 2009). Implicitly the assumption is that the industrialized countries can invest in CERs wherever it is cheapest globally, leading to a decrease in overall compliance costs (UNFCCC 2011).

Under the protocol, developing countries do not have an emissions cap (Nazifi 2010). CDM projects can earn CERs for both private and public sectors. The term "projects" refers to any Annex I "sponsored" activity or undertaking in a developing country that serves to decrease overall carbon emission from the baseline. An example of such project includes "rural electrification project using solar panels or the installation of more energy-efficient boilers" (UNFCCC 2011).

The entire mechanism is overseen by the CDM Executive Board, which operates under Conference of the Parties (COP/MOP) of the United Nations Framework Convention on Climate Change (UNFCCC), and is available only to those countries that have ratified the Kyoto Protocol (UNFCCC 2011; Nazifi 2010). In order for a project to qualify for transferable credits, the Designated National Authorities (DNA) must approve it. The DNA is an individual or group of individuals in a specific country that has been designated to approve CDM projects (UNFCCC 2011). In order for a project to be approved, it must create emission reductions that are less than those of the current state, i.e., business-as-usual (UNFCCC 2011). The process for administering carbon credits for CDM projects contains seven steps: (1) project design and formulation; (2) national approval; (3) validation and registration; (4) project financing; (5) monitoring; (6) verification and certification; and (7) issuance of CERs (Nazifi 2010). The Executive Board can issue the CERs once the CDM projects complete the registration process and can begin generating credits from the starting date of the project activity (Nazifi 2010; Zhang *et al.* 2009).

Teräväinen (2009) explains the two objectives of the Clean Development Mechanism as being a mechanism for industrialized counties with emission reduction commitments to assist developing countries in (1) implementing their own reduction projects and thereby, earning certified emission reduction credits (CERs) and (2) supporting their sustainable development. Teräväinen (2009) discusses the trade-offs that have made it difficult for industrialized countries to support both objectives, especially the latter regarding sustainable development, as it is understood by the Brundtland definition. It highlights, however, the success, especially in Finland, of cross-sectoral collaboration in an attempt to gain CERs. Teräväinen (2009) uses four themes of an eco-modernist approach to assess the success of CDM in distinguishing the concept that economic growth inherently leads to environmental degradation; promoting clean technology; integrating environmental and other governmental policies; and utilizing alternative, economic policy approaches for environmental issues. Referencing Finland's CDM policies, Teräväinen (2009) explains how in order to integrate environmental and climate change policy to enable the transfer of emission reduction technologies to developing countries and remain aligned with international climate policy, multiple national departments concerning technology, the environment, economics, development, energy, etc. as well as international actors, have created a mutual dependence upon one another. Through this integration, Finland has held steadfast that through innovative technology creation, they have grown economically without cost to the environment and have been able to take advantage of clean technology as a successful business export (Teräväinen 2009).

Although using an eco-modernist lens to analyze the CDM recognizes major achievements in certain policy areas, Teräväinen (2009) acknowledges how it lacks the objective of sustainable development in developing countries. In the case of the CDM, when gaining CERs through diffusing technologies is the priority for industrialized countries, they lose sight of the broader objectives of sustainable development. Teräväinen states that the initiatives are designed to lower developing countries' emissions, with the focus on gaining CERs; however, the initiative evolves into an economic incentive to see a project through the most cost-effective manner as possible where they can most cheaply implement it, rather than where the most environmental improvements and social assistance are needed. In doing so, it is also not cost-effective to address the main local development needs of the community and use the time and resources for a bottom-up approach to see that the project contributes to the sustainable development of the host country in reality. Teräväinen (2009) acknowledges that this creates a highly inequitable usage of the CDM, as project selection becomes so highly selective in order to locate the most efficient usage of donor countries' resources (Teräväinen, 2009).

While Teräväinen seems to promote many of the domestic policy actions that have been put into play to achieve the CDM, these actions illustrate the political problems of scale with such policy. By turning environmental action over to market forces by linking it with economic incentives, the ease of ignoring the real problems and potential risks for developing countries is too great. The motivation to address a global problem by working at a local level is erased because the international market for carbon and technology holds more power than the needs of developing countries. Even if, ideologically, countries are trying to do good, market forces do not allow them to assign values to the sustainability of their projects and the contribution to host countries. And unfortunately the ideology of many countries is rather transparently to displace the blame of their carbon emissions, continue with their business-as-usual, and concern themselves with safe and inexpensive ways to offset their emissions. Their motives can be concealed by virtuous rhetoric in their initiatives, but as evidenced, the rules surrounding such initiatives do not uphold their promises. This ideology may not be as apparent or meaningful to developing countries who see the CDM as an opportunity for investment and knowledge/technology transfer in order to speed their growth. The knowledge gap created by industrialized countries seems to provide false expectations of a commitment to sustainable development in a host country, but often results in benefiting the investor country more.

Dinar *et al.* (2011) also acknowledge the apparent success of the CDM in that through such development assistance, by 2012 it will have reduced CO_2 emissions by 800 to 1150 million tons in the form of CERs. They assume, however, that the initiative is evolving into a form of foreign direct investment, whereas Teräväinen contended as well that investors are merely looking for the lowest opportunity cost, thereby eliminating many developing countries from receiving assistance. Their study finds that with thirty-four investor countries and 175 potential host countries, there are 5950 plausible pairs for the CDM to be enacted, yet only 305 of those pairs through 5669 projects have been at work together. In an attempt to study the factors that have led to existing dyads of participating countries, the authors evaluate countries on the measure of the colonial ties between two countries, the membership of host countries to international governmental organizations, the energy use and source of that energy in host countries, as well as vulnerability of governance, and ease of doing business within the host country. Several findings were concluded to increase the likelihood of collaboration: countries with strong colonial ties or trade relationships; democratic and demilitarized governments; and involvement in international governing institutions which all contribute to the ease of doing business with the country (Dinar *et al.* 2011). Also worthy of noting is the energy use of both countries involved, as the CDM is only efficiently utilized should a country find that it is more cost-effective to lower emissions elsewhere rather than in their own industries, and many least developed countries do not have the industry and power sectors to pollute enough for the CDM to earn an investor country enough CERs (Lederer 2011).

Many are surprised and inspired by the volume of CDM projects, but the rules of the CDM have prevented the development of many meaningful projects and allowed for the creation of some projects completely unaligned with its goals. Martello and Dargusch (2010) explain how impossible/difficult it has been to initiate afforestation/reforestation projects, as it took two years for the rules regarding them to be established, and the complexity of them makes it a very unviable option. Padma (2011) revealed that in India government projects using renewable energy do not qualify to receive CERs, yet private initiatives using mostly coal power have been approved. India has been plagued by countless incidences where public development projects are displaced/ruined by private initiatives that do not actually contribute to sustainable development that is devoted to improving society and the environment. Padma claims that while the rules of the CDM have prevented many potentially deserved CERs, they have allowed for the granting of many credits with little investment, and the cost of CERs is already crashing with little transition to low-carbon economies to show for it. Recently, however, projects amounting to $1.08 billion were approved to help eight developing countries in a variety of mechanisms including clean energy technology transfer, forest management, and low-carbon development (Hurtado 2011). The participation of locals was cited as being a huge success, and the representation of countries assisted is promising. More than 2400 CDM projects have been registered and have an expected annual average of 383 million certified emissions reductions (CERs) in sixty-nine countries (Duan 2011).

The following specific challenges have been widely discussed in designing and implementing CDM projects:

1   The criteria for additionality and establishment of baseline are inadequate.
2   There is unclear empirical evidence on whether true reductions in global net emissions are taking place.
3   CDM projects' foci are disproportionately focused on "clean energy."
4   There is inequitable displacement of responsibility.

A major component of an approved CDM project is that the planned reductions would not occur without the additional incentive provided by emission reduction credits. This is a concept referred to in the literature as "additionality" (Zhang *et al.* 2009; Nazifi 2010). Due to the international and multi-tiered market of CDM projects, additionality is difficult to prove. This is problematic for a variety of reasons. While proponents of CDM may argue that a transfer of funds to developing nations is not necessarily a loss, the main concern with proving additionality is that it backs up the entire approval process for CDM projects (Norman 2011). It takes three years, on average, for a CDM project to be approved and assigned CER credits (Bryan *et al.* 2010). This creates a "bottle-neck" in the market, which discourages both host countries and suppliers from participating in the mechanism (Sutter and Parreño 2007).

Similarly problematic is the inability to establish a real baseline comparison for these projects. This is a critical component of CDM because CER credits are awarded based on the difference between the emissions reductions resulting from the project and the baseline emission (Zhang *et al.* 2009). If the baseline is wrongly established, especially if it is reported lower than it is, then the CER buyer's emissions are not actually being offset, which could lead to an actual increase in overall carbon dioxide emissions instead of the perceived decrease (Clean Development Mechanism 2011).

Proponents of the Clean Development Mechanism (Zhang 2011), assert that, based on 2009 figures, it is expected that the 4200 CDM projects will decrease global greenhouse gas emissions by 2900 million metric tons for carbon dioxide equivalents by the end of 2012. If that figure is accurate, that is comparative to 40 percent of the carbon emissions coming from the United States in 2007 (Zhang and Wang 2011). However, because of the flaws in CDM project regulation, we have no way of knowing just how accurate that figure is. Sutter and Parreño (2007) examined a number of different CDM projects around the world. The study showed only 72 percent of projects were actually decreasing their carbon emissions on the same level that CERs were being exchanged for them. And even if Zhang's figures are accurate, impressive as they are they are not enough to slow climate change. This is perhaps more due to the fact that the Kyoto emission targets themselves were set too low to begin with. That is to say that even if all countries that ratified the Kyoto Treaty *did* stay within their contracted emission caps, the decrease in emissions is predictably not enough to slow global warming to less than two degrees celcius (above pre-industrialization levels), a goal set by climatologist who say that is the temperature increase where we will start seeing, or continue to see worse, ecologic impacts (Sawyer 2011). For others, the 2°C goal might be too lax and calls for more stringent targets have been issued, such as 350 ppm target (McKibben 2011)

Clean Development Mechanism projects are expected to also meet the Kyoto objective of contributing to sustainable development (Zhang *et al.* 2009; UNFCCC 2011). Projects are considered "sustainable" largely due to the fact that projects aimed at reducing greenhouse gases will yield environmental co-benefits such as decrease sulfur dioxide emissions (Zhang *et al.* 2009). Unfortunately, a study by Sutter and Parreño (2007) suggests that as few as 1 percent of CDM projects are contributing to sustainable development. This could be due to the fact that the Kyoto Protocol only vaguely defines "sustainable development," even though it is one of its key goals (Sutter and Parreño 2007; Zhang *et al.* 2009). The UNFCCC states that it is up to the non-Annex I countries, the host countries, to define their own "sustainable development" requirements (Sutter and Parreño 2007). Therefore, the "dual-success" of CDM must again be called into question.

In 2008, an overwhelming 82 percent of CDM projects were geared towards "clean energy," including renewable energy, fuel switching, and energy efficiency projects (Nazifi 2010). As of 2009, nearly 50 percent of CDM projects were focused solely on renewable energy (Bryan *et al.* 2010). Meanwhile, only 1 percent of CDM projects for 2009 were focused on afforestation/reforestation efforts. Additionally, CDM does not explicitly prevent deforestation or degradation (Kinderman *et al.* 2008; Zhang *et al.* 2009).

The CDM program has been fraught with corruption and has experienced limited success in prompting a proliferation of sustainable technologies in the developing world (Cox 2010: 118). Additionally, the CDM's current structure does not incentivize greenhouse gas reductions below target levels. Schneider (2009: 95) argues that the CDM is a "zero game to the atmosphere," meaning that any certified reduction in a developing nation can be met with an equal rise in emissions in the corresponding developed nation if it is already meeting its reduction timeline.

Apparently, the CDM has two objectives: to secure funding for sustainable development initiatives in industrializing countries (with the goal of leapfrogging fossil fuel technologies); and to allow developed nations to meet greenhouse gas targets in an economically efficient manner. However, within the current organization of the CDM, these two goals are sometimes pitted against each other. Sutter and Parreño observe that

> most [developing] countries will not have the market power to considerably influence the global market price for emission reductions. Competition among parties in attracting CDM investments could create an incentive to set low sustainable development standards in order to attract more projects with low abatement costs.
>
> (Sutter and Parreño 2007: 76)

In a sample of CDM initiatives worldwide, Sutter and Parreño found that most projects contributed only to one of the policy's objectives and of these, 95 percent of the volume of CERs issued were related to products that ranked low on an index of sustainable development (2007: 89).

Because CDM projects are only monetized in terms of their total abatement of greenhouse gas emissions, the true value of projects that promote sustainable development not directly related to GHG mitigation is not realized in market pricing of CERs. The sustainable development projects evaluated on a longer temporal scale will ultimately reduce the severity of pressures and consequences of continued growth. Therefore, any pricing of CERs must include considerations of the future benefits of a sustainable development project as they relate to reducing the vulnerability of future generations to climate change. This would necessitate the formation of an index or ranking system, in which a development can be evaluated based upon its environmental and social benefits, particularly poverty easement. An alternative to this more complex pricing mechanism would be to add a premium value to CERs related to high sustainable development projects, which "might increase the share of such projects in global carbon markets" (Sutter and Parreño 2007: 89).

The legitimacy of CDM also faces several operational hurdles. As previously mentioned, its current design does not incentivize the reduction of emissions below target levels. Second, the carbon credit market has been extremely volatile, which led to a crash in 2006 and an outflow of investors (MacDonald 2010: 18). A simple solution that could address both these flaws would be to establish a discounting rate for CERs, so that whatever carbon credit value they are assigned, it will be lower than the actual amount of emission reduced by a project. This is effectively akin to a carbon credit retirement scheme, which reduces the market pool of CERs, necessitating more mitigation projects. The advantage, however, with a CER discounting scheme is that it can be scaled in relation to sustainable development criteria. For example, a project that ranks highly in sustainable development criteria would experience less of a CER discount than a project that ranks lower, incentivizing development projects that are more relevant for developing nations. This would also have a positive effect on the carbon market. By introducing a discounting system, the number of CERs that are issued will be decreased, resulting in more consistent price increases and restoring appeal to investors (Schneider 2009: 108). In this manner, a discounting scheme for CERs could serve to reconcile the dual objectives of the CDM, by promoting high-value sustainable development projects, stabilizing the carbon market, and mitigating greenhouse gas emissions.

From the politics of scale perspective, the Clean Development Mechanism (CDM) is a policy mechanism set into place to create a global solution for what has traditionally been seen as national or state problems in international development and energy sectors. The local and national scale politics, for example, demonstrate that CDM projects can be used on many levels and manipulated in many ways. Repetto (2001: 327) identifies several potential flaws within the CDM framework. One such flaw he calls the "Incentive Incompatibility" and describes as,

> Stripped to the basics, CDM is an international payment for a service, the reduction of greenhouse gas emissions. A fundamental flaw is that neither the seller nor the buyer of the credit has a private interest in the actual delivery of the service.
>
> (Repetto 2001: 327)

In the framework of the CDM, a project gets emission credits for greenhouse gas emissions that extend past what would have existed without the implementation of this project. Kevin Smith identifies this problem as "additionality" and argues that "there are some studies coming out saying some 30–50 percent of projects under the CDM are non-additional. That is, it's just business-as-usual that's having a CDM label tagged on it in order to provide extra financing" (WhatProductionsUK 2008).

De Soto Blass (2010) proposes to build upon the CDM in the post-Kyoto timeframe by including either a resource rent extraction or a Pigouvian Tax on the negative externalities produced in the development of any project applying for the CDM label. In this way, some larger environmental issues could be included with the CDM policy mechanism.

The CDM has also been criticized by participating countries for its low efficiency, complicated and inapplicable methodological requirements, lack of transparency, imbalanced regional distribution, and uncertain international rules after 2012 (Duan 2011). The additionality is key to the validity of the CDM projects because without it the impact, and subsequent emission reductions of the donor countries contributions would be unknown. Yet, this additionality is difficult to prove and some officials have expressed that they feel the process is arbitrary (Schiermeier 2011). Each project also has to demonstrate the maintenance of environmental integrity. The CDM has to ensure each of these requirements on a trade-by-trade basis and this process requires a lot of time and administrative resources leading to high transaction costs (Boyle *et al.* 2009). In 2008 it took between 150–250 days to complete the registration process, and in 2010 the gap between monitoring and issuances of credits was up to ten months (Duan 2011). The design of the system has created a trade-off between transaction costs and credibility. Fewer regulations verifying the validity of the projects, or standardization of registration would reduce these costs and increase efficiency but jeopardize the credibility of the CDM.

The greatest weakness to CDM and to all climate change mitigation efforts is the ideological commitment to promote a market oriented solution that controls the flow of resources from global north to local south. These issues are discussed as marketization of climate governance in the latter chapters.

With the expiry of the first commitment period of the Kyoto Protocol in 2012 drawing closer, there are now various discussions on the future design of the CDM. For example, in 2008 a Civic Exchange meeting was held in Hong Kong to discuss the effectiveness of sustainable development mechanisms such as CDM. During that meeting, Gail Kendall, a member of the Civic Exchange, talked about the shortfalls of the CDM. She stated,

> The whole of the targets of the Kyoto Protocol are not really at the scale that we would hope in terms of climate stabilization … the whole market for Clean Development Mechanism is not at a scale of the growth in China and India … it's helpful in a portion, but its not really going to address the total growth in emissions in countries such as China and India; it's just not at that scale at the present time.
>
> (Kendall 2008)

She went on to add that,

> a relatively small amount of the Clean Development Mechanism is actually renewable energy…. I checked the UN site yesterday and added up that only just over 15 percent of the total CDM certificates were for renewable projects like wind power, hydro power, biomass, etc. … and this is because low carbon energy cost more.
>
> (Kendall 2008)

Lund (2010) argued that,

> During the last few years, the supervisory system of the Kyoto Protocol's Clean Development Mechanism has been heavily criticized for not being able to guarantee the additionality of CDM project activities. The two main themes in the recent critique of the CDM have been concerns about the additionality rules being unclear, and doubts about the objectivity of the DOEs when applying these rules due to their profit-seeking behavior. But although these critiques have been useful in highlighting problems in the CDM, the ensuing analysis of how to solve these dilemmas has so far not been well grounded in theory. Therefore there is a lack of understanding of the root causes to these problems.
>
> (Lund 2010: 278)

In other words, there is a huge problem in the delegation system of the CDM; it doesn't take into account the concepts of the actual theory of delegation that was introduced in institutional economics in the 1970s. However, the critics of CDM are very narrow-minded in the respect that there are adjustments that can be made to fix the problem. To make delegation work better in the supervisory system of the CDM, two fundamental changes are necessary. First, the problem of the Designated Operational Entities' competing loyalties has to be resolved. It is not credible that project developers select and pay the Designated Operational Entities that are supposed to supervise their projects. It has been suggested that the independence of the Designated Operational Entities could be strengthened if the CDM Executive Board or the UNFCCC Secretariat selected and paid the Designated Operational Entities. This suggestion goes to one of the root cause of the CDM's current problems, which is that agents have an economic incentive to be more loyal to third parties than to the principal. If the CDM Executive Board as principal were to select and pay the Designated Operational Entities, agents who follow its preferences could be favored, which would thus create a system of carrots and sticks (Lund 2010). Such a system could be financed through the share of proceeds levied on CDM projects to cover administrative costs. Even if the share of proceeds had to be increased to cover the expenses, this arrangement would at worst be cost-neutral to project developers, as they would no longer have to engage the Designated Operational Entities themselves. The second proposed change in a post-Kyoto regime is regarding the objectivity of the additionality rules—basically, they need to be increased. Even if Designated Operational Entities were selected and paid by the CDM Executive Board, the current principles used for determining additionality are very vague and leave much room for interpretation, which means that the application of rules will necessarily be arbitrary. Several suggestions on how to increase the objectivity of additionality testing are currently circulating in the climate policy community. Promising options discussed in the negotiations include a positive or negative list of project activity types that are to be considered additional by definition, the use of dynamic benchmarks, and the development of conservative, standardized baselines for specific project types to be used for additionality determination. Using four such standardized principles for additionality testing could form the basis for a less arbitrary additionality determination practice, without requiring excessively detailed regulation from the CDM Executive Board. Even if this would not solve all the problems of the CDM's supervisory system, combined with a reform that eliminates the perverse incentives for the Designated Operational Entities to be more loyal to project developers than to the CDM Executive Board it would at least move the system closer towards a healthy balance in the inevitable trade-off between monitoring needs and specialization gains (Lund 2010).

As Gail Kendall brought up in the meeting and as many skeptics argue, CDM isn't fully beneficial because of its low contribution to sustainable development, unbalanced regional and sectoral distribution of projects, and its limited contribution to global emission reductions. More specifically, the major concerns about the functioning of the CDM are that it has often been noted that the CDM does not lead to global emission reductions but is, at best, a mechanism that offsets emission increases in industrialized countries. The contribution of the CDM to sustainable development in the host countries is widely seen as being very limited. Asian and Latin American countries make up more than 95 percent of the projects and certified emissions reductions (CERs) in the CDM project pipeline, raising concerns about the equitable regional and sub-regional distribution of projects). The unequal distribution of CDM projects among sectors has also been noted, with the transport and building sectors, both key for achieving ambitious climate targets, almost completely absent from the project portfolio. Certain project types, notably the destruction of industrial gases such as HFC-23 and N, are thought to generate high windfall profits for project developers and host countries. Projects face significant transaction costs due to the institutional and governance structure of the CDM.

Even despite these concerns, however, there are changes that can be implemented in the CDM in the post-Kyoto timeframe that could allow for better results. First, preferential treatment for under-represented host countries or preferable project types appears to be an option without significant negative impacts. And even though it most likely won't significantly change the current sectoral and regional distribution, it could still provide modest support to some countries and project types presently bypassed by the CDM at little political cost (Bakker *et al.* 2011). Second, minimum thresholds for sustainable development set at the international level, and verified by Designated Operational Entities, may improve the sustainable development profile of the CDM project portfolio (Bakker *et al.* 2011). Third, differentiation based on quotas or eligibility of countries or project types could significantly change the regional distribution of CDM projects. The supply of credits would probably be reduced under these options, but would be sufficient to meet most 2020 demand scenarios (Bakker *et al.* 2011). And lastly, CER discounting could contribute positively to most of the issues, in particular by creating a mechanism that results in net global GHG emission reductions (if the discount rates are higher than the share of non-additional projects being registered in the system). Also the discount rates can be applied and adjusted so that they benefit under-represented countries in the CDM market or benefit project types with a particularly strong contribution to sustainable development. The discounting of appropriate project types may equally reduce windfall profits as well. And even though there are definite drawbacks for these CDM recommendations, the benefits outweigh the costs, especially considering the lack of expected results over the last ten years. One more policy recommendation for the CDM could be the idea of having an international standard, i.e., a taxonomy for measuring and monitoring sustainable development benefits. Defining sustainable development isn't an easy task. The SD benefits of CDM projects can be defined and accounted for in numerous ways. For good reasons, therefore, no single, authoritative definition exists of how to assign unambiguous criteria and indicators covering all aspects of sustainability. The innovativeness of the taxonomy is to assess the sustainability of CDM projects in a simple, qualitative way and present findings at aggregated levels rather than at the project level (Olsen and Fenhann 2008). The most important policy implication of the taxonomy is its contribution towards a new verification protocol to ensure that potential sustainable development benefits of CDM projects are actually realized

The fact of the matter is that the carbon credits system is designed to assist the developing countries to fund otherwise unaffordable emissions reduction projects and to undertake renewable energy initiatives. This should enable them to build more sustainable, less carbon-intensive economies, driven increasingly by the generation of renewable energy. A company or a public entity such as a municipality can therefore undertake a GHG emissions reduction project and offset some, and possibly all, the costs by selling CERs, provided that the project meets CDM requirements. The CDM can therefore be seen as a novel means of attracting foreign investment to the developing countries. For example, If a Chinese mine cuts its methane emissions under the Clean Development Mechanism, and receives financial incentives to do so, then there is a real global climate benefit. Likewise, if windmills are installed in Morocco and accelerate the implementation of that country's clean energy policy or make its goal more achievable, then there is a clear global benefit for the climate. The polluter who buys the credits from the offset has contributed to reducing global emissions in a verifiable manner. It matters little to the global climate where an emission reduction has taken place; what's important is that emissions overall are being reduced. Carbon offsetting is therefore not a zero-sum game, but a highly effective tool to reduce global emissions. Despite the cost effectiveness of CDM, the sheer reductions in global scale cumulative GHGs from the CDM projects remain a very marginal, small scale, contribution that should be a big cause of concern for post-Kyoto climate governance regime designers. A solid mitigation set of policy actions will be needed in the developed countries, which are afflicted with the politics of ideology, as discussed in the next chapter.

4 The politics of ideology I

Risk perceptions and psychology of denial

This chapter provides an in-depth application of the politics of ideology lens, with an emphasis on explaining the interactions of ideology on the dynamic formation of risk perceptions of citizens and policy-makers in the developed countries with a focus on the United States. The politics of ideology lens is applied to explain the psychology of denial that is observed in many climate change risk assessment studies. Further, ideologically driven political gridlock, driven by a variety of politicians, corporations, mass media and citizens, is analyzed from the perspective of a politics of ideology lens. Mental and cultural model elicitation techniques are used to evaluate the systematic effects of ideology on climate change deniers and sceptics.

Assessment of risk perceptions in complex societies

Human induced climate change is arguably one of today's most important scientific and social challenges (McCarthy *et al.* 2001). To effectively address the challenge of global climate change, some scholars argue that it is "imperative to have an accurately and completely informed public" (Trumbo 1996: 281) and that improvements in public understanding are urgently needed (see NSF 1999; Moser and Dilling 2007). The U.S. government asserts that a public that can "accurately interpret complex scientific information" will make informed decisions about how to reduce the risk of climate change (U.S. Department of State 2002: 149). Internationally, it has been argued that public knowledge of climate change is critical to public interest, and translating concern for climate change into action requires public knowledge (see Bord *et al.* 2000: 205; Stern 2008: 33). It has been suggested that public knowledge of climate change science is critical to developing public health policy (Frumkin *et al.* 2008), and to shape public understanding of economic analysis (Stern 2008: 1). Some researchers assert that public understanding of climate change science is "the most powerful predictor of both stated intentions to take voluntary actions and to vote on hypothetical referenda to enact new government policies to reduce greenhouse gas emissions" (Bord *et al.* 2000: 205). Furthermore, such proponents argue, mitigation policies require public acceptance for effective implementation and therefore must take public values into account in order to secure the highest levels of public participation: "If a problem and the actions people can take to help solve it are framed in ways that resonate with cultural values and beliefs, people are more likely to take the action than if they are not" (Moser and Dilling 2004: 41).

Such arguments for improving public understanding of climate change science are built on the premise that if the public has accurate scientific knowledge then they will make public policy decisions that fit in with the scientific normative view. This premise leads numerous scholars to conclude that "the scientific community should devote greater resources to developing public understanding of these principles to provide a sound basis for assessment of climate policy proposals" (Sterman and Sweeney 2007: 229).

If public understanding of climate science influences public acceptance of climate policy, then empirical studies of factors that influence public interpretation of science can assist in developing effective risk communication strategies, which in turn can influence the development of effective mitigation and adaptation policies. Toward this end, Bostrom *et al.* (1994) deployed a mental modeling approach to assess public understanding of climate change science. They found that respondents' "explanations of the physical mechanisms underlying global climate change were inconsistent and incomplete." Many respondents were found to hold other fundamental misconceptions and more subtle misperceptions (Bostrom *et al.* 1994: 968). The authors concluded that the respondents'

> flawed mental models restricted their ability to distinguish between effective and ineffective strategies. One particular concern is that laypeople may waste their energies on ineffective actions, such as conscientiously refusing to use spray cans, while neglecting such critical strategies as energy conservation.
>
> (Bostrom *et al.* 1994: 969)

These findings, and others (Kempton 1991, 1997), suggest that "correcting" mental and/or cultural models about climate change science is a necessary step for developing effective mitigation and adaptation strategies.

This conclusion about the need to correct mental and/or cultural models of public knowledge of climate change science, however, poses daunting problems for research on public understanding of science. These include the problems of "cognitive dissonance" (Stoll-Kleemann *et al.* 2001), "trust-gap" hypothesis (Priest *et al.* 2003) and ideology (Carvalho 2007; Wood and Vetlitz 2007), all of which seem to suggest that flaws in the mental or cultural models of climate change may not be as easy to "correct" as argued by the proponents of the seamless improvability of public understanding of climate change science by educating the public about anthropogenic climate change (Bostrom *et al.* 1994; Trumbo 1996; U.S. Department of State 2002; Sterman and Sweeney 2007)

Stoll-Kleemann *et al.* evaluated schematic patterns in social psychology to understand "cognitive dissonance," which explains both the encoding and the retrieval of information are often guided by personal desire to maintain cognitive consistency. The lack of consistency is the state of the *dissonance*. In general, individuals experiencing dissonance seek to resolve it, deny it, or displace it.

(Stoll-Kleemann *et al.* 2001: 111, see also Taylor and Fisk 1981)

Their research demonstrates that, "for the most part, denial or displacement act powerfully to maintain the gap between attitude and behavior with regard to climate change norms." They conclude that there is "both a coherence and a rationality to dissonance and denial that will not make it easy for democracies to gain early consent for tough climate change mitigation measures" (Stoll-Kleemann *et al.* 2001: 115). Cognitive dissonance, or the psychology of denial presents difficulties to shifting mental or cultural models through mere infusion of climate change science.

Similar arguments have been made about the lack of trust in the government and the trust gap between environmentalists and industrialists in the context of biotechnology policy development (Priest *et al.* 2003). The trust gap hypothesis can also be extended to climate change policy development in the sense that the public may perceive that climate change knowledge is politicized before its infusion in public minds. Vast uncertainties about climate change impacts make it even likelier that climate change science is politicized, which can be trusted by one social group and distrusted by another social group. The lack of trust in the climate change scientific knowledge may make it difficult to establish the "norm" for correcting the mental or cultural models of public.

Current discussions of public understanding of science suggest that factors such as cognitive dissonance and trust gap hypothesis pose challenges in communicating science to public audiences. These findings suggest we must take into account public attitudes to understand how citizens will interpret scientific data. Zia and Todd (2010) evaluate how and to what extent ideological beliefs influence public understanding of and concern for climate change in hopes of gaining a greater understanding of the underlying cultural models of this globally significant issue.

Ideology influences the interpretation and eventually the understanding of scientific predictions such as climate change forecasts. "Ideology works as a powerful selection device in deciding what is scientific news, i.e., what the relevant 'facts' are, and who are the authorized 'agents of definition' of science matters" (Carvalho 2007: 223). While there are multiple definitions of ideology (e.g., Eagleton 1991), Zia and Todd (2010) treat ideology, proposed by Carvalho,

as a system of values, norms and political preferences, linked to a program of action vis-à-vis a given social and political order. People relate to each other and to the world on the basis of value judgments, ideas about how things should be, and preferred forms of governance of the world. In other words, ideologies are axiological, normative and political.

(Carvalho 2007: 225)

Value judgments and social norms constitute ideological preferences that influence individual actions in response to the social and political order.

Fundamentally, ideology influences knowledge. Wood and Vedlitz (2007), for example, demonstrated in a nation-wide survey (N = 1093) of U.S. citizens that political ideology systematically influenced the concern of citizens along the lines of the much acclaimed "guns versus butter" political agenda theory (Mintz and Huang 1991; Antonakis 1999; Clark 2001; Carrubba and Singh 2004). Early studies (Russett 1969, 1970; Hartman 1973; Wilensky, 1974) of the guns versus butter trade-off postulated that government spending in defense and security (guns issues) detracts from the spending in education, health and social welfare (butter issues). Some of the later studies (Caputo 1975; Clayton 1976; Peroff and Podolak-Warren 1979) found no direct evidence for this trade-off, while others (Mintz and Huang 1991) found indirect evidence, i.e., increased levels of military expenditures dampen investment, which reduces growth, thereby reducing the ability of governments to allocate more funds to social welfare programs. This macro-economic theory of guns versus butter trade-offs was extended to studying the effects of ideology on public opinion formation (Carrubba and Singh 2004) and political agenda formation (Wood and Vedlitz 2007). The extended guns versus butter analogy postulates that citizens with a liberal political ideology tend to be more concerned about "butter" issues; while citizens with a conservative political ideology tend to be more concerned about "guns" issues. This analogy is useful to understand public perceptions of climate change, which is typically framed as a "butter" issue, because of its domestic effects on economy and environment.

In their study of public attitudes toward "guns" and "butter" issues, Wood and Vedlitz (2007) used linear regression models to test the effects of various independent variables such as the political ideology, religious identity, party affiliation, socio-demographic background, income, age, and gender on the concern of citizens regarding five important public policy "guns" and "butter" issues: terrorism, economy, health care, poverty and global warming. They found that the concern of citizens for these five issues is significantly affected by political ideology, as predicted by the "guns versus butter" theory. They also found the effects of party affiliation, religious identity, race, gender, college education, general science literacy, and self-reported knowledge of the issue to be significant; while income, age, number of children and number of information sources had no generally significant effect on citizen concern for these five issues. As a butter issue, one that is not explicitly linked to national defense, global warming was found to be of higher concern for citizens with liberal ideologies while conservatives ranked guns issues, such as terrorism, as having greater importance. This raises the question of how reframing climate change as a "guns" issue might broaden public concern for the issue.

A closer look at Wood and Vedlitz's (2007: 558–561) results reveals an underlying pattern of conflicting effects on citizen concern for global warming. While conservative political ideology has a negative effect on citizen concern for global warming, factors such as college education, global warming literacy, personal experience with global warming, attention to global warming and social network concern about global warming have either insignificant or positive effects on citizen concern for global warming.

Climate change, ideological divide, and the psychology of denial

Zia and Todd (2010) build on Wood and Vedlitz's model and investigate the interplay of positive and negative effects of political ideology and other socio-demographic and economic variables on citizen concern for global warming. They postulate that ideology has a significant effect on public understanding of climate change science and test the proposition that *even college education does not trump the significant effect of political ideology on citizen understanding of and concern for climate change problem.* They argue that if their proposition tests positive, it will imply that enhanced educational opportunities, such as media sources, may improve communication of climate change, but improved education of citizens alone will not reduce misunderstanding of climate change science caused by ideological preferences. They use mental and cultural models approach to demonstrate the effects of ideology vis-à-vis college education on public understanding of climate change science and discuss different strategies for improving climate change science communication across ideological divides.

Our cultural models are influenced by the complexities of climate change as a scientific and more importantly political issue. Scientific pronouncements on issues that concern the health and safety of the general public have high news value. Media coverage can attract public attention (Weingart *et al.* 2000: 280) and can alter public perceptions of risk. "The degree of importance and concern attributed to climate change does indeed fluctuate in relation to other events and newsworthy items" (Lorenzoni and Pidgeon 2006: 75). Public perceptions of environmental issues are influenced by scientific and technical knowledge as well as a variety of social factors including values and worldviews (see Slovic 2000; Leiserowitz 2006). For these reasons, social acceptance of climate policy depends on ideology, as individuals and groups make decisions on adaptation strategies based on political beliefs and socio-cultural experience.

Citizen concern about different political, economic and social issues can lead to the formulation and development of public policies. In addition to climate change mitigation, the effectiveness of climate change adaptation policies depends on "social acceptability of options for adaptation, the institutional constraints on adaptation" (Adger 2003: 388). Public beliefs about the science of climate change constrain government implementation of "policies consistent with the best available scientific knowledge" (Sterman and Sweeney 2007: 230). Reducing greenhouse gas emissions to levels needed to mitigate future climate change requires "changes in technology, energy prices, business practices, consumer behavior, and other activities affecting people's daily lives" (Sterman and Sweeney 2007: 233). For effective levels of participation in mitigation practices, people must understand the scientific basis for proposed policies: when "the best available science" conflicts with existing mental models that form the basis for common-sense approaches, people are "unlikely to adopt appropriate policies or generate political support for legislation to implement them" (Sterman and Sweeney 2007: 233).

Effective policy thus requires understanding what influences public perception of climate change science and public assessment of responsibility. Representation of science to the public, such as in media and government outlets, is important for "assessing the responsibility of both governments and the public in addressing climate change" (Carvalho 2007: 223). Public opinion polls highlight a sense of urgency associated with climate change as an environmental issue, but not necessarily as a domestic issue (Lorenzoni and Pidgeon 2006: 80). Thus, U.S. citizens may not feel a sense of responsibility to take action on a global issue. Public attitudes toward climate change are complex: individuals characterize climate change in multiple terms "related to their everyday experiences and locality, distinguishing effects on different scales in space and time" (Lorenzoni and Pidgeon 2006: 80). Widespread public "buy in" requires communication built on existing mental models that citizens already used to make decisions about participating in public policy. This calls for further study of communication of science in the context of existing mental models, such as those that construct political ideological frameworks.

Mental models can be represented as complex networks of beliefs that are dynamically contingent upon the place and time of a citizen (Axelrod 1976; Kosko 1986; Zaltman 1997; Morgan *et al.* 2002). Concept mapping techniques can be used to represent mental models and, furthermore, shared mental models, i.e., cultural models, can be elicited by tracking the shared patterns of belief networks (Christensen and Olson 2002). Zia and Todd (2010) build upon this mental/cultural modeling research and postulate that public understanding of science is a function of such models, which are dynamically evolving belief networks about ideology, religious identity and other socio-demographic and economic factors.

More specifically, they hypothesize that citizens adhere to cultural models that exhibit patterns of belief networks that systematically differ across various gradations of political ideology, religious identity, socio-demographic background, income and sources of public/media information. Further, they postulate that citizens, for a given place and time, exhibit mental models with patterns of belief networks that can be differentiated across a series of ideological preferences. More specifically, they test following three specific hypotheses in this study:

[H$_1$] Ideology has a significant effect on citizen concern for global warming. As citizens' ideology shifts from liberal to conservative, concern for global warming decreases.

[H$_2$] Citizens with college education and higher general science literacy tend to have higher concern for global warming.

[H$_3$] College education does not increase global warming concern for ideologically predisposed conservatives. In other words, the effect of ideology trumps the effect of college education when it comes to being concerned about global warming.

To explore the proposed ideologically driven cultural models and testing corresponding hypotheses, Zia and Todd (2010) replicated the survey questionnaire of Wood and Vedlitz (2007) on a local scale by surveying residents of California's San Francisco Bay Area in spring 2008. They interviewed 655 Bay Area adult residents to represent the population of ~6.9 million San Francisco Bay Area residents.

In this survey, variables about present and future concern regarding five issues—terrorism, health care, economy, global warming and poverty—are measured on a scale from 0 (no concern) to 10 (maximum concern). These variables about citizen concern are used as dependent variables in the regression models. Political ideology is measured on a seven-point scale from strongly liberal (1) to strongly conservative (7). Figure 4.1a shows the variation in the citizen concern about five most important present issues by three gradations of political ideology: liberal, moderate and conservative. The liberals are less concerned about terrorism (a guns issue), while conservatives are less concerned about global warming (a butter issue). Similar patterns emerged for issues of future concern, as shown in Figure 4.1b.

Responses reveal 16.76 percent of the sample affiliates with the Republican Party. The average income of the sample is ~$60K (below the Bay Area's median household income of ~$76K), and the average age is ~32 years (below the Bay Area's median age of ~38 years). Approximately 59 percent of respondents in the sample identified as white, which is very similar to the Bay Area's estimated 57.2 percent white population in 2008. The sample has 47 percent females, which is ~2.9 percent less than the Bay Area's estimated female proportion in 2008. About 34 percent of sample respondents report having one or more children. Conservatives represent 15 percent of the sample. There are ~40 percent college graduates in the sample. This mirrors the rate of the area's population, 41 percent of those 25 years or older have a bachelor's degree or higher. Figure 4.2 shows the variation by educational background of respondents with their reported present concern about five most important issues. Respondents with college education (tertiary category in Figure 4.2) are less concerned about terrorism, while those with primary school education only are less concerned about global warming.

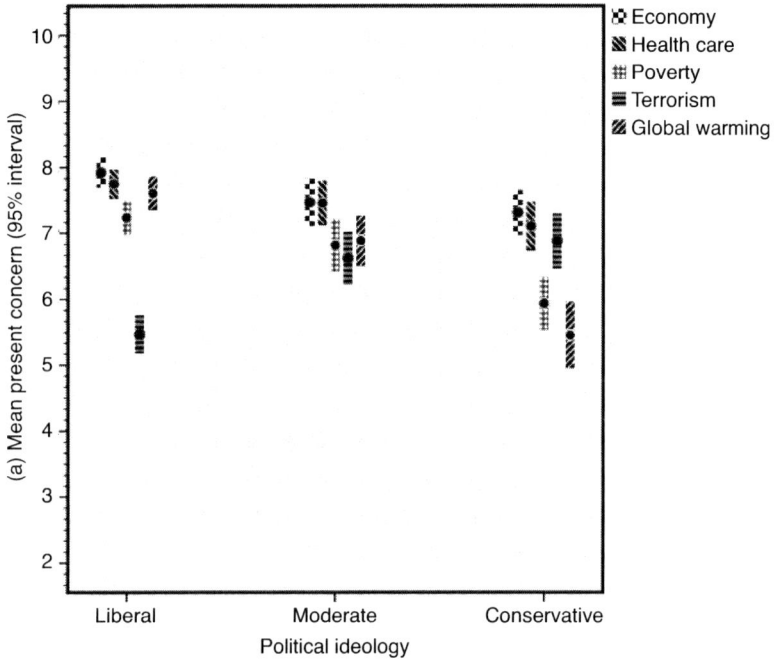

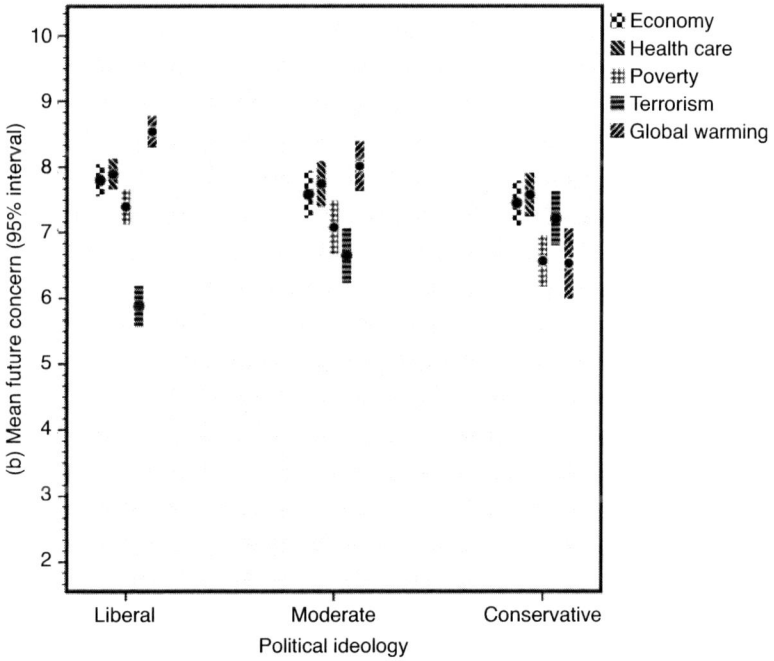

Figure 4.1 (a) Present and (b) future concern about economy, health care, poverty, terrorism, and global warming distributed by political ideology of the respondents.

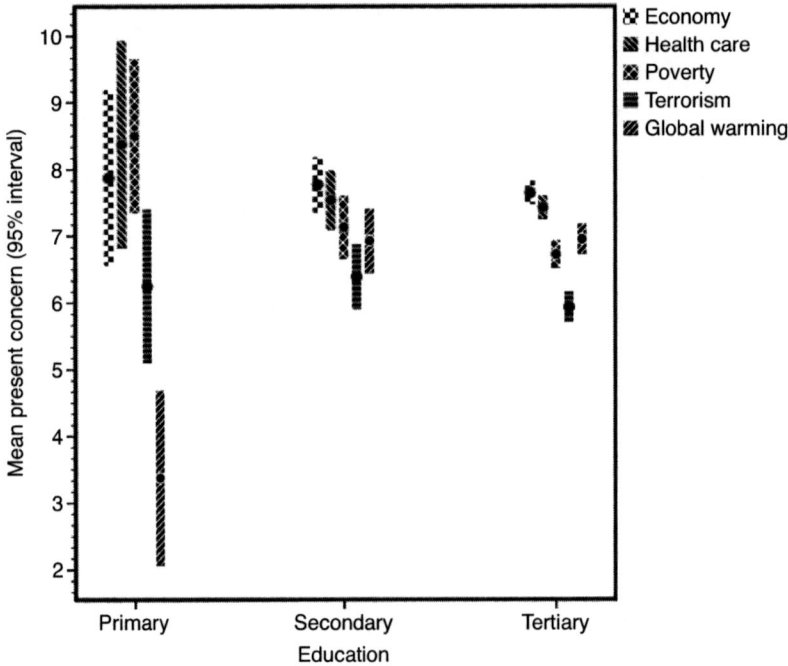

Figure 4.2 Present concern about economy, health care, poverty, terrorism, and global warming distributed by educational qualification of the respondents.

Four questions measure general science literacy in the survey, such as electrons are smaller than atoms (true or false); the oxygen you breathe is from plants (true or false); the sun goes around the earth (true or false); and human beings developed from earlier species of animals (true or false). These four questions are selected from a larger set of questions that have been typically asked to respondents in Eurobarometer and other similar projects for measuring their scientific literacy (Durant *et al.* 1989; Evans and Durant 1995). Out of these four general science literacy questions, Zia and Todd (2010) found that only about 40 percent of the sample was able to correctly answer all four questions. The variable "general science literacy" measures the correct number of answers to these four questions. On a scale of 1 to 4, corresponding to correctly answered questions, the sample mean of general science literacy stands at 3.08 ± 0.91. These four questions are a relatively crude measure of general science literacy and they have significant methodological limitations in capturing the scientific literacy of the survey respondents, as extensively discussed by Pardo and Calvo (2002, 2004).

Similarly, three questions measure global warming literacy, such as: nitrous oxide is a greenhouse gas (true or false); the major cause of increased atmospheric concentration of greenhouse gases is human burning of fossil fuels (true or false); and aerosols are airborne particles that are known to contribute to the formation of clouds and precipitation (true or false). Only 20 percent of the sample was correctly able to answer all three questions. The variable "global warming literacy" measures the correct number of answers to these three questions. On a scale of 1 to 3, corresponding to correctly answered questions, the sample mean of global warming literacy stands at $1.54 + -0.99$. Again, global warming literacy is a crude measure of quantifying citizen literacy on climate change issues and it suffers from similar methodological limitations that apply to Eurobarometer type of questions derived for quantifying scientific literacy (e.g., Pardo and Calvo 2002, 2004).

Linear regressions models were derived by Zia and Todd (2010) to estimate significant variations in the present and future concern of citizens for the five most important issues. Table 4.1 shows results of five regression models with present and future concern for five issues as the dependent variables. The independent variables in the regression models of Table 4.1 are the same as used by Wood and Vedlitz (2007: 558–559) to analyze the nationwide survey sample in their study. They find their statistical model fairly robust and confirm the broader guns versus butter political agenda theory for their study sample.

More specifically, Wood and Vedlitz (2007: 558) found that as the ideology of respondents moves from liberal to conservative on the seven-point scale, respondents' present concern for terrorism significantly increases by 0.19 points, comparable with 0.18 point increase in their sample, as shown in Table 4.1. Wood and Vedlitz (2007: 558) report 0.26 and 0.48 point less concern among conservative ideologues for poverty and global warming, respectively. Zia and Todd (2010) sampled conservative ideologues have similarly 0.23 and 0.28 point less concern for poverty and global warming, respectively. Respondent concern about future concern for these five issues is also similar across Wood and Vedlitz's (2007: 559) and Zia and Todd's (2010) study samples (Table 4.1). Wood and Vedlitz (2007:559) find a 0.45 point less future concern for global warming among conservative ideologues, while Zia and Todd (2010) find a 0.32 point less future concern for global warming among conservative ideologues.

Zia and Todd (2010) are able to confirm their first hypothesis [H_1] that ideology does have a significant effect on citizen concern for global warming. Further, as citizens' ideology becomes more conservative, concern for global warming significantly decreases.

In terms of testing their second hypothesis [H_2], their sample results are different from those reported by Wood and Vedlitz (2007: 558–559). Wood and Vedlitz found that college education significantly reduces present citizen concern for terrorism, economy, health care and global warming by 0.41, 0.38, 0.34 and 0.48 points respectively. Zia and Todd's (2010) study results in Table 4.1 show that college education has no significant effect on reduced concern for these five issues. In terms of future concern (Table 4.1), their study does show that college education significantly reduces citizen concern for global warming by 0.49 points (compared with insignificant effect in Wood and Vedlitz model).

Table 4.1 Determinants of level of concern for five important issues facing the nation now and in the future: reduced models

Determinant condition	Terrorism		Economy		Health care		Poverty		Global warming	
	Now	Future	Now	Future	Now	Future	Now	Future	Now	Future
Ideology	0.18*	0.18*	−0.18*	−0.10	−0.07	−0.10	−0.23**	−0.14	−0.28**	−0.32**
	(0.09)	(0.09)	(0.07)	(0.07)	(0.07)	(0.07)	(0.07)	(0.08)	(0.08)	(0.08)
Party Republican	0.88*	1.16**	−0.24	−0.62*	−0.52	−0.26	−0.51	−0.51	−1.05**	−1.12**
	(0.36)	(0.37)	(0.30)	(0.31)	(0.28)	(0.28)	(0.30)	(0.31)	(0.31)	(0.33)
Income	−0.001	−0.03	0.04	0.08**	−0.01	0.04	0.02	0.01	0.01	0.04
	(0.03)	(0.03)	(0.03)	(0.03)	(0.03)	(0.03)	(0.03)	(0.03)	(0.03)	(0.03)
Age	0.02*	0.01	−0.002	0.003	0.01	0.005	−0.008	−0.001	−0.005	−0.007
	(0.009)	(0.01)	(0.008)	(0.008)	(0.008)	(0.008)	(0.008)	(0.008)	(0.008)	(0.009)
Religious conservative	0.55	0.22	0.51	0.45	−0.01	0.18	0.006	0.23	−0.36	0.01
	(0.36)	(0.38)	(0.31)	(0.32)	(0.29)	(0.29)	(0.32)	(0.32)	(0.33)	(0.34)
Race: white	−0.03	0.17	−0.38	−0.03	−0.11	−0.26	−0.02	0.02	0.34	−0.05
	(0.24)	(0.25)	(0.20)	(0.21)	(0.19)	(0.20)	(0.21)	(0.21)	(0.22)	(0.23)
Gender: female	0.61*	0.74*	0.27	0.58*	0.61**	0.85**	0.80**	0.85**	0.84**	0.87**
	(0.28)	(0.29)	(0.24)	(0.24)	(0.22)	(0.22)	(0.24)	(0.24)	(0.25)	(0.26)
Children	0.01	−0.47	−0.42	−0.14	−0.47	−0.09	0.03	0.07	−0.21	−0.15
	(0.34)	(0.35)	(0.29)	(0.30)	(0.27)	(0.27)	(0.29)	(0.30)	(0.30)	(0.31)
Gender Female* Children	−0.16	0.55	0.42	0.23	0.50	0.19	0.10	0.23	0.44	0.31
	(0.49)	(0.51)	(0.42)	(0.43)	(0.39)	(0.40)	(0.42)	(0.43)	(0.43)	(0.45)

Education: college	−0.26	−0.22	0.14	0.01	−0.002	−0.20	−0.22	−0.20	−0.43	−0.49*
	(0.25)	(0.26)	(0.21)	(0.22)	(0.20)	(0.20)	(0.22)	(0.22)	(0.23)	(0.24)
General science literacy	−0.27*	−0.17	0.02	−0.06	0.02	−0.11	−0.13	−0.05	0.06	0.04
	(0.13)	(0.14)	(0.11)	(0.12)	(0.11)	(0.11)	(0.12)	(0.12)	(0.12)	(0.13)
Number of information sources	−0.15	−0.17	0.11	0.15	0.03	−0.05	−0.06	−0.09	−0.005	−0.05
	(0.11)	(0.11)	(0.09)	(0.10)	(0.09)	(0.09)	(0.09)	(0.10)	(0.10)	(0.10)
Self-reported knowledge of issue	0.37**	0.29**	0.09*	0.07	0.28**	0.19**	0.49**	0.44**	0.42**	0.30**
	(0.04)	(0.05)	(0.04)	(0.04)	(0.04)	(0.04)	(0.04)	(0.04)	(0.04)	(0.05)
Constant	3.17**	3.97**	7.22**	6.55**	5.51**	6.76**	5.14**	4.91**	4.87**	6.98**
	(0.69)	(0.72)	(0.60)	(.62)	(0.56)	(0.57)	(0.62)	(0.63)	(0.63)	(0.65)
N	495	495	496	496	495	495	496	496	494	493
σe average	2.53	2.63	2.15	2.22	2.03	2.05	2.20	2.23	2.26	2.36
Adjusted R²	19.3%	14.0%	2.7%	3.9%	12.3%	8.2%	28.6%	23.0%	29.7%	22.7%
F	10.06**	7.18**	2.06*	2.55*	6.35**	4.37**	16.22**	12.37**	17.01**	12.12**

Notes

* shows significance at 0.05 level and ** shows significance at 0.001 level. Numbers in brackets show standard errors. F statistic shows joint significance for all variables.

Similarly, in terms of testing the effect of general science literacy on present citizen concern, Wood and Vedlitz (2007: 558) found that general science literacy significantly reduces citizen concern for terrorism (0.22 points), economy (0.26 points) and health care (0.18 point), but general science literacy has no significant effect on citizen concern for poverty or global warming. Zia and Todd (2010) study results show no significant effects of general science literacy on global warming, poverty, health care or economy, but higher general science literacy does seem to significantly reduce citizen concern for terrorism (0.27 points).

Overall, they reject hypothesis #2, as they are not able to confirm whether college education and general science literacy increases citizen concern for global warming. This hypothesis will require further testing in future studies that use more sophisticated questions to assess scientific literacy, as recommended by Pardo and Calvo (2004). One possible explanation for the counter-intuitive evidence that college education decreases future citizen concern for global warming may be the enhanced belief in the capacity of human societies to adapt to change. This explanation will also require further empirical testing.

For testing their hypothesis #3 about the interaction effect of college education and political ideology, they re-specified the above regression models by adding an interaction variable (college education*ideology) in the list of independent variables. They also added five additional variables—global warming literacy, science unclear, personal experience, attention and social network concerned—to further improve explanatory power of the regression models. The re-specified regression models reported in Table 4.2 have in general higher R^2 and lower model error as compared to regression models reported in Table 4.1. Wood and Vedlitz have not tested the effect of college education*ideology in their regression models, but they do report the results of five additional variables just for present and future global warming concern (2007: 561).

Table 4.2 shows the results of re-specified regression models. Zia and Todd (2010) reject the null hypothesis with 95 percent confidence level that college educated conservative ideologues have similar concern about global warming (both present and future) as non-college educated conservative ideologues. The interaction variable is statistically significant in the negative direction in Table 4.2 for the regression equations with present and future concern for global warming as dependent variable. This implies that college educated conservative ideologues have lower concern for global warming both in present (0.26 points) and future (0.29 points). College education does not appear to trump ideology of their sampled bay area residents, presumably one of the most progressive area within the United States, the largest cumulative emitter of GHGs at the global scale.

Unlike Wood and Vetlitz (2007: 561), Zia and Todd (2010) found no significant effect of global warming literacy or personal attention on increased concern for global warming. Conversely, Wood and Vedlitz (2007: 561) found no significant effect of social network concerned on present or future citizen concern about global warming, but their study found a significant effect of social networks for present (0.15 points) citizen concern about global warming. The effect of social networks on future concern for global warming is, however, insignificant in their study. While Wood and Vedlitz (2007: 561) found that "science unclear" perception of citizens reduces global warming present (0.22 points) and future (0.17 points) concern, they found a stronger effect of "unclear science" in the same direction for both present (0.42 points) and future (0.63 points) concern of citizens about global warming.

Implementation of this survey reveals complex interaction effects among variables exerting positive and negative effects on citizen concern for various policy issues. Zia and Todd (2010) analyze this interaction complexity by constructing and comparing cultural models of citizens with different political and religious ideologies. They use survey data of San Francisco Bay Area residents to construct a universe of cultural maps, a simplified version of which is shown in Figure 4.3. Statistically significant correlations (in + or—percent points) show the corresponding magnitude and direction of positive and negative relationships among important variables shown as nodes in Figure 4.3. This simplified figure shows only 20 percent or higher correlations among the study variables. Dark grey and light grey connect the nodes. Dark grey lines show negative correlations and light grey lines show positive correlations. Lines are scaled according to their magnitudes, i.e., Pearson correlation coefficients.

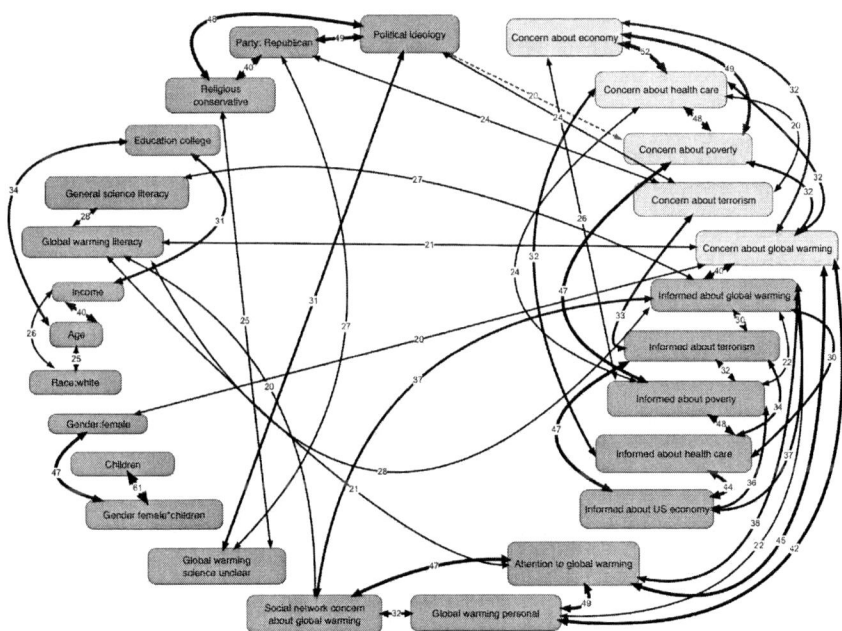

Figure 4.3 Universe of cultural models (numbers represent significant Pearson Correlations, 20% or higher).

Table 4.2 Determinants of level of concern for five important issues facing the nation now and in the future: expanded models

Determinant condition	Terrorism		Economy		Health care		Poverty		Global warming	
	Now	Future	Now	Future	Now	Future	Now	Future	Now	Future
Ideology	0.10	0.11	-0.15	-0.10	0.05	-0.03	-0.07	-0.06	-0.05	-0.05
	(0.12)	(0.12)	(0.10)	(0.10)	(0.09)	(0.09)	(0.10)	(0.10)	(0.09)	(0.10)
Party: Republican	0.47	0.81*	-0.47	-0.70*	-0.97**	-0.56	-0.47	-0.37	-0.75*	-1.12**
	(0.39)	(0.40)	(0.32)	(0.34)	(0.30)	(0.31)	(0.33)	(0.33)	(0.31)	(0.33)
Income	0.01	-0.04	0.03	0.09**	-0.01	0.06*	0.03	0.03	0.02	0.03
	(0.04)	(0.04)	(0.03)	(0.03)	(0.03)	(0.03)	(0.03)	(0.03)	(0.03)	(0.03)
Age	0.02*	0.02	-0.001	0.000	0.01	0.000	-0.009	-0.003	-0.01	-0.01
	(0.01)	(0.01)	(0.009)	(0.009)	(0.008)	(0.009)	(0.009)	(0.009)	(0.009)	(0.009)
Religious conservative	0.56	0.10	0.47	0.39	-0.12	0.02	0.007	0.27	-0.47	-0.01
	(0.40)	(0.41)	(0.33)	(0.35)	(0.31)	(0.32)	(0.34)	(0.34)	(0.32)	(0.34)
Race: white	-0.23	0.13	-0.33	-0.04	-0.11	-0.31	-0.03	0.05	0.38	0.02
	(0.27)	(0.27)	(0.22)	(0.24)	(0.21)	(0.22)	(0.23)	(0.23)	(0.21)	(0.23)
Gender: female	0.60*	0.84*	0.37	0.67*	0.73**	1.01**	0.98**	0.81**	0.67**	0.87**
	(0.30)	(0.31)	(0.25)	(0.27)	(0.23)	(0.24)	(0.26)	(0.26)	(0.24)	(0.26)
Children	0.08	-0.30	-0.30	-0.09	-0.27	0.04	0.25	0.27	-0.33	-0.24
	(0.37)	(0.38)	(0.31)	(0.33)	(0.29)	(0.30)	(0.32)	(0.34)	(0.30)	(0.32)
Gender female* children	-0.25	0.48	0.20	-0.01	0.21	-0.10	-0.19	0.12	0.72	0.43
	(0.54)	(0.56)	(0.45)	(0.48)	(0.42)	(0.44)	(0.46)	(0.47)	(0.43)	(0.47)
Education: college	-1.03	-0.96	0.20	0.10	0.50	0.19	0.70	0.44	0.42	0.54
	(0.59)	(0.61)	(0.49)	(0.52)	(0.46)	(0.48)	(0.50)	(0.51)	(0.47)	(0.51)
General science literacy	-0.28	-0.14	0.000	-0.06	0.02	-0.10	-0.19	-0.03	-0.16	-0.17
	(0.16)	(0.16)	(0.13)	(0.12)	(0.12)	(0.13)	(0.13)	(0.13)	(0.13)	(0.14)

Number of information sources	-0.15 (0.12)	-0.16 (0.12)	0.15 (0.10)	0.21* (0.10)	0.14 (0.09)	0.05 (0.10)	0.02 (0.10)	-0.03 (0.10)	0.04 (0.10)	-0.03 (0.10)
Self-reported knowledge of issue	0.35** (0.05)	0.33** (0.05)	0.13** (0.05)	0.09 (0.05)	0.33** (0.04)	0.22** (0.04)	0.51** (0.04)	0.47** (0.04)	0.26** (0.05)	0.20** (0.05)
Global warming literacy	-0.01 (0.13)	0.03 (0.13)	0.14 (0.11)	0.14 (0.11)	0.10 (0.10)	-0.04 (0.11)	-0.10 (0.11)	-0.20 (0.11)	0.18 (0.11)	0.21 (0.11)
Global warming science unclear	0.09 (0.19)	-0.008 (0.19)	0.15 (0.16)	0.04 (0.17)	0.29* (0.15)	0.15 (0.15)	-0.22 (0.16)	-0.07 (0.16)	-0.42** (0.15)	-0.63** (0.16)
Personal experience with global warming	0.10 (0.18)	0.000 (0.18)	-0.16 (0.15)	-0.07 (0.15)	0.03 (0.14)	-0.14 (0.14)	-0.03 (0.15)	-0.09 (0.15)	0.46** (0.14)	0.26 (0.15)
Attention to global warming	-0.47* (0.20)	-0.29 (0.21)	0.02 (0.17)	-0.06 (0.18)	-0.02 (0.16)	0.10 (0.16)	0.19 (0.17)	0.28 (0.17)	0.30 (0.16)	0.27 (0.18)
Social network concerned about global warming	0.07 (0.05)	0.04 (0.06)	0.04 (0.04)	0.03 (0.05)	-0.03 (0.04)	-0.03 (0.04)	-0.06 (0.05)	0.000 (0.05)	0.15** (0.04)	0.04 (0.05)
Ideology*education college	0.22 (0.15)	0.22 (0.15)	-0.03 (0.12)	-0.03 (0.13)	-0.18 (0.11)	-0.13 (0.12)	-0.27* (0.13)	-0.18 (0.13)	-0.26* (0.12)	-0.29* (0.13)
Constant	4.08** (1.09)	4.13** (1.12)	6.61** (0.90)	6.65** (.96)	4.08** (0.85)	6.16** (0.88)	4.99** (0.93)	4.15** (0.94)	3.68** (0.85)	6.89** (0.91)
N	420	421	421	421	420	420	421	421	419	419
α average	2.51	2.59	2.10	2.22	1.97	2.04	2.15	2.17	2.02	2.17
Adjusted R^2	18.0%	14.7%	3.2%	3.9%	16.1%	9.6%	30.3%	24.9%	42.9%	33.6%
F	5.85**	4.80**	1.73*	1.90*	5.24**	3.34**	10.61**	8.35**	17.55**	12.14**

Notes
* shows significance at 0.05 level and ** shows significance at 0.001 level. Numbers in brackets show standard errors. F statistic shows joint significance for all variables.

Figure 4.3 shows that the cultural models of politically conservative ideologues and religious conservatives demonstrate ideological preferences that are systematically different from the cultural models of liberal ideologues and religious liberals (or even moderates). While females in general appear to be more concerned about every public issue, younger people appear to be more liberal and older people appear to be more conservative. While educated people have relatively higher general science and global warming literacy, as well as higher incomes, college education does not significantly correlate with religious or political ideologies. Further, it is not educational background, but religious and political ideologies that influence citizen concern about policy issues such as global warming, terrorism, poverty, health care and economy. College education is, however, strongly correlated with higher income, age and race.

The assessment of cultural models, as shown in Figure 4.3, demonstrates the strength of correlation effect of ideology, religious identity and political party affiliation vis-à-vis other background variables on public concern about global warming. This implies that citizen cultural models are driven by ideological predispositions, which have strong effects on citizen understanding of scientific forecasts, while college education has no significant effect in shaping ideologically driven cultural models. Improving the education of citizens will thus be not enough to communicate the urgency and implications of climate change science. More sophisticated strategies will need to be developed to communicate climate change forecasts that cut across ideologically divided cultural models.

If education alone cannot overcome the politics of ideology, what could be the driving motif of a post-Kyoto climate governance and policy regime? There are two divergent responses to this question. On the one hand, we have "framing" proponents, who propose "re-framing" the climate change science for policy-makers and the general public that cuts across ideological divide and "reaches" the climate deniers and sceptics. On the other hand, we have "institutional" proponents, who suggest that this ideological divide is more of an "institutional" question. Since right-wing political and religious ideologues are committed to protecting the institutional ideology of "free markets," "capitalism" and "minimum governmental regulation," they view climate change policy and governance regimes as ultimately violating these longstanding ideological utopias. Both of these "framing" and "institutional" approaches act as antidote to the persistent politics of ideology. Next, I present the political gridlock that this ideological divide has presented to the world as a gift of modern capitalism.

Ideology and political gridlock

Currently in the world, many developed nations have come so far and become so successful in terms of economic growth and global trade relations, and yet, most of the rest of the world's developing countries and regions have been left behind, and now face not only the problem of maintaining economic stability and trade relations with the developed countries, but also the daunting challenge of global climate change and global warming. Global climate change, despite its speculators, has worsened over the last 300 years and has now indeed becoming an environmental disaster for Planet Earth, a disaster that is negatively affecting the planet and all of the species that live on it. In the last 200 to 300 years, all of the developed countries around the world, especially the United States, have been operating under the idea of economic growth and economic development; the more people, the more money, and the more growth the better. However, the problem with this business-as-usual approach of unlimited growth is that there is little to no regard to the well-being and safety of the earth's environment, and as a result we are paying for it as global climate change accelerates and as the earth's environment deteriorates more and more every single day. And perhaps the most troubling aspect of global warming and global climate change in addition to its environmental costs is its distribution of disproportionate negative effects across regions of the earth—or, in other words, the harsh consequences of unlimited growth and global climate change have a much greater negative impact on the developing world than on the already developed world—continents such as Africa and South America are being hit hard by the negative effects of global climate change and yet they don't have the money or the resources to do much about it. And as the Kyoto Protocol comes to a conclusion, we all ask ourselves the all-too-familiar question, "What are we going to do next in order to save this planet and its environment?" The familiar answer is "nothing."

This "nothing" is obvious from the lack of any meaningful action in major GHG emitting countries (with perhaps a few exceptions, e.g., Germany and Scandinavia). If we review the history of any proposed legislation in the U.S. Congress and EU, it becomes obvious that the right-wing parties in both the United States and EU vehemently oppose policy proposals aimed at mitigating global climate change. Worse still, some of the right-wing commentators even go to the extent of equating climate policy with neo-socialism. Consider, for example, Bell's (Larry Bell: Climate of Corruption) statement that climate change "has little to do with the state of the environment and much to do with shackling capitalism and transforming the American way of life in the interests of global wealth redistribution." If one happens to visit the websites of right-wing think tanks such as the American Heartland Institute, Competitive Enterprise Institute, and CATO institute, the politics of ideology displayed against climate change and global warming manifests in the form of attacks on climate change scientists and any policy-makers who accept the fundamental premises of anthropogenic climate change. Eventually, this politics of ideology translates into a coherent movement that is persistently launched against any proposed climate change mitigation policies, such as Waxman-Markey Bill in the United States. While it is well known that this vehement politics of ideology is used to block climate change legislation at the federal level, the worst part of this political drama is that this war is funded by the oil and coal lobbies, who will be the primary losers if a meaningful climate policy is enacted in the United States and European Union.

Notwithstanding the failure at the national level in the case of the United States (and continental level in the case of the European Union), recently there have been some "optimists" in the climate change governance community who have argued that climate change policies have been more meaningfully embraced at the local and state levels in the United States. California's adoption of the Global Warming Solutions Act (AB 32) is given as one example. The formation of ICLEI is given as another example. While these efforts at the state and the local level are commendable, many independent evaluation studies of these local to state level actions have found very insignificant effects of GHG reductions. Further, if there were any effects of these local and state level policies, they must show up in the aggregate U.S. and EU level GHG data. However, as will be discussed in later chapters, the GHG emissions do not seem to be abated by these local and state level actions.

While it is perhaps early to pass a judgment on California's actions, or the carbon trading markets established in the EU and the north-eastern United States, the aggregate GHG emissions in both the United States and the EU continue to rise. At the core of this dilemma is the underlying economic system that generates GHGs as an externality. Further, I will argue that the environmentalists and left-leaning parties in both the United States and the EU have mistakenly tried to propose that climate change solutions could be found within the "existing" economic and political systems. Every proposed solution entails trade-offs, and winners and losers. Even the imposition of Pigovian taxes, e.g., carbon tax or a carbon cap and trade policy regime, will not be adequate to the deal with the magnitude of the climate change mitigation problem that we are facing as a human civilization. At best, carbon taxes and/ or cap and trade markets could signal energy and transportation sectors to shift their emission profiles. These policies will entail upfront costs that right-wing political ideologues construe as an attack on capitalism and left-wing political ideologues construe as unpleasant for present generations in terms of employment loss in coal and oil industries, at the minimum.

If we undertake a careful analysis of the GHG reductions that are needed to meet 350 ppm or even 450 ppm CO_2 stabilization target, I will demonstrate in Chapters 6 and 7 on the politics of knowledge that Pigovian taxes in energy and transportation sectors at the global scale are not going to be adequate. The anthropogenic global climate change problem is much larger than the energy and transportation sectors. It pervades human land use, deforestation, large-scale manufacturing, and heavy industries. In a nutshell, the anthropogenic global climate change is the ultimate curse of modern capitalism that has defined human civilization since the dawn of industrial revolution. Ironically, right-wing political ideologues have been more astute to judge this than the environmentalists and left-wing political ideologues, as the latter have so far mistakenly signalled that a simple carbon tax or cap and trade policy should be "good enough" to "deal" with the climate change problematic. Not surprisingly, we are in the midst of a persistent political gridlock that defines the failure of Kyoto Protocol in arresting the growth of GHGs at the global scale, or even the failure of UNFCCC process so far in negotiating a post-Kyoto climate treaty, as seen since the UNFCCC COP negotiations in Copenhagen in 2009. Will "reframing" the uncertain climate change risk solve this civilizational scale problem? Next, in Chapter 5, I discuss this reframing in the context of communicating risk and uncertainty to ideological disparate populations and interest groups.

5 The politics of ideology II

Communicating uncertain climate change risk

Coping with the politics of ideology

Building upon the analysis in the previous chapter, this chapter provides an over-view of coping with the politics of ideology both through reframing climate science communications of risky and uncertain information, which is inherent in scientific forecasts, and through focusing on institutional reforms at local, regional, national and international scales. First, the challenges of communicating uncertain climate change risk are described in the larger context of behav-ioural psychology and experimental decision-making research. Next, some specific proposals to reframe climate science communication are addressed and compared with re-institutionalization challenges.

The challenges of communicating risk and uncertainty

The literature on risk communication and risk perception makes a compelling argument that individuals cannot be described as rational economic actors when it comes to assessing and communicating scientific information containing risk and uncertainty. There are a variety of decision heuristics that prevent individu-als from undertaking purely rational decision-making processes when faced with risk and uncertainty as is the typical case with the local to regional scale climate change mitigation and adaptation policies. These decision heuristics influence the way that individuals compare ambiguous and uncertain risk information (Fox and Tversky, 1995). While there is an extensive research going on in neuro-sciences, experimental decision-making, and behavioural psychology about identifying such decision heuristics, here I briefly provide a review of the fol-lowing heuristics in the context of climate change risk communication: affect heuristic, risk setting, ambiguity and uncertainty aversion, trust-deficit, framing problem, representative heuristic, and the availability heuristic; all of which sin-gularly or combined severely affect the uncertain climate change risk communi-cation from the climate change scientists to policy-makers and the general public. Not surprisingly, many of these decision heuristics are effectively wielded by the powerful interest groups to distort the scientific messages and muddy the climate science, which in turn perpetuates the politics of ideology in coping with anthropogenic climate change. These decision heuristics also discredit the linear model of the science–policy interface, which assumes that the objective science could be used to inform the public policy. Given the complex trade-offs and large-scale winner–loser effects that are entailed in any climate change policy, a straightforward communication of uncertain climate change risk information is not going to be simply possible.

Affect heuristic

An "affect heuristic" causes individuals to skew their risk perceptions according to general positive or negative associations that individuals have with a certain source of risk (Slovic *et al.* 2004). In addition, risk perceptions are also influenced by the setting in which risk information is being communicated, as some settings and communicators have a history of deceitful risk communication (Weiss *et al.* 1995). Evidence for the affect heuristic and risk setting issues is provided by Leiserowitz's (2006) study of the impact that decision heuristics have on climate change risk perception.

Slovic *et al.* (2004) provide a theoretical background for the idea that decision-making is a mix between analytical and experiential frames of mind, and supports the theories using experimental empirical findings. Experiential decision-making is defined as being guided by "affect," which are positive or negative associations that individuals have with certain images. Depending on the individual and the actions being considered, the reliance on affect for decision-making varies. After reviewing neurological and theoretical support for the affect heuristic, the authors examine empirical studies that support its importance.

First, Slovic *et al.* (2004) describe research on the "dread" heuristic, which shows that individuals' perception of riskiness is increased if they dread the outcome. In other words, individuals perceive the risk of a nuclear power plant as greater than the risk of a car crash, despite the fact that the statistical probability that the latter will occur is much greater. This is because the image of a nuclear disaster is more dramatic and provokes a "dread" affect.

Slovic *et al.* (2004) cite research on the relationship between risk and benefit in risk perception, demonstrating that if individuals have a positive view of the benefit of an activity they are more likely to underestimate the risk. Conversely, if individuals see as activity as being low risk, they are more likely to overestimate the benefit. Slovic *et al.* (2004) also argue that the way a risk is presented can have an impact on the extent to which decision-making is influenced by affect, as risks described using narrative led to much higher risk perception than risks described using percentages.

Slovic *et al.* (2004) demonstrate how proportion dominance influences risk perception, citing research that found that airline passengers are more likely to support a safety investment if the benefit is framed as saving the lives of 98 percent, 95 percent, or 90 percent of 150 passengers, rather than being phrased simply as saving the lives of 150 passengers. Slovic *et al.* (2004) argue that this study demonstrates the affect heuristic, as proportion gives decision-makers a reference point, which enhances their ability to appreciate the meaning of an outcome. However, Slovic *et al.* (2004) also found that the impact of proportion diminishes when there is a clearer affect attached to a decision. For example, the probability associated with winning the lottery has a negligible impact on respondents' positive association with the lottery.

Slovic *et al.* (2004) also examine some of the limitations of taking an experiential approach to decision-making. The limitations are caused by deliberate manipulation of affective associations by advertising and by the existence of limits of the experiential system to adequately assign affects to stimuli. Thus, risks that are distant in time and space are harder to appreciate, particularly if a person has never experienced a similar risk. Slovic *et al.* (2004) reference studies of young cigarette smokers who fail to appreciate the risk of nicotine addiction until they are already addicted. Cigarette smoking is thus an example of both limitations of experiential decision-making, as individuals are manipulated into applying a positive affect to the act of smoking, while at the same time their capacity to develop a sufficient negative affect for nicotine addiction is diminished by their lack of experiential knowledge and the time distance of the risk.

Slovic *et al.* (2004) argue that the risk can be better managed by harnessing the knowledge about the limitations of the experiential and analytical approaches to risk perception. They conclude that experiential thinking can help individuals better appreciate risks that are commonly expressed in an analytical format.

Leiss (1995) examines similar issues in the realms of environmental and health risk communication. Leiss (1995) argues that in the setting of health and environmental risk communication, there is a history of distrust, and as a result parties enter risk assessments in such domains with a large degree of skepticism as they assume that risk communicators have an incentive to exaggerate risk assessments. Leiss (1995) uses the analogy of a poker game, in which participants distrust each other because they are aware that there is an incentive, and even expectation, that players will be deceitful. Leiss (1995) provides several examples of strategic, deceitful behavior among both environmentalists and industry advocates. For example, Leiss (1995) reviews a case study involving environmentalist and labor union concerns about chemicals used in the lumber industry. The case study illustrates several examples of participants bluffing other parties by making exaggerated claims about risks. Thus, in a setting characterized as a deceitful game, participants are more likely to distrust the risk assessments provided by actors. Leiss's (1995) study demonstrates the limitations of the expert affect concept identified by Fox and Tversky, as it demonstrates that risk communications can be provided by a variety of experts who are not equally trusted (or trusted at all) by risk assessors.

Leiserowitz (2006) examines how risk assessments of global warming have developed among the broader population. Leiserowitz (2006) used a representative mail in survey to measure respondents' risk perception, holistic affect, affective imagery, cultural values, policy preferences, and demographic characteristics. Risk perception was measured by asking respondents to rank the degree of risk that global warming posed to different areas and populations. Holistic affect was derived by asking respondents to rank the degree of their positive or negative feelings towards global warming. Affective imagery was measured by conducting a word association game in which respondents listed associations with climate change and then suggested whether or not they had positive or negative feelings towards the association. To measure cultural values, the respondents were asked to answer 25 questions, as the responses were categorized according to an individualism and hierarchism index. Respondents were also asked to indicate their level of support for different policy alternatives for addressing climate change, and their responses were coded into a policy preferences index and a tax policy preferences index. After collecting the data, Leiserowitz (2006) tested to see if there was a statistical correlation between affect imagery and values on risk perception and policy preferences.

The Leiserowitz (2006) study found that respondents' affective imagery in relation to climate change predominately involved distant risks such as melting ice caps and negative impacts on non-human nature, and such images were correlated with lower levels of risk perception. Some respondents were identified as having an "alarmist" affect, while others were identified as having a "naysayer" affect. The alarmist and dry/desert affect were associated with high levels of perceived risk, while the naysayer, politics, and don't know affects were associated with low risk perception. The Leiserowitz (2006) study also found that individualism and hierarchism were associated with low risk perception, while egalitarianism was associated with high risk perception. Females, minorities, liberals, members of environmental groups and newspaper readers were demographics that all had high risk perceptions, while whites, males, conservatives, and registered voters all low risk perceptions. When comparing all the predictors, the holistic affect had the strongest correlation, the "naysayer" affective image had the second strongest, negative image affect the third, and egalitarianism the fourth. This shows that affect, imagery, and values are stronger predictors than socio-demographic variables. The same correlations were also found in relation to national policy support, with naysayers, holistic affect, and egalitarianism being the three strongest predictors. For tax policy preferences, there was a correlation with both imagery and values. For both policy preferences, values had the greatest correlation.

Consistent with both Slovic *et al.* and Weiss, the Leiserowitz (2006) study demonstrates that the meaning attached to both information about risks and the risks themselves has an impact on individuals risk perception. The high levels of risk perception associated with alarmist images, such as world destruction, demonstrates the dread affect and also the affective image that can be developed by a narrative description of the risk. The naysayer and politics affect can both be characterized as a distrust of information sources that are provided in the environmental risk realm. The finding that the affective imagery of distant global warming impacts was associated with low risk perception is aligned with Fox and Tversky's finding (described below) that ambiguity aversion in relation to geographically near phenomenon is typically higher in comparison than ambiguity aversion in relation to a geographically distant phenomenon.

Ambiguity and uncertainty aversion

Fox and Tversky (1995) ran six experimental studies on university students (and in some cases faculty) to assess whether or not comparative ignorance can limit the influence that ambiguity aversion has on risk assessments. In the first three studies, participants were presented with two bets, one with a certain probability and one with an uncertain probability. Next, the studies tested what would happen if participants were only exposed to one bet, half of which having a certain probability and the other half having an uncertain probability. In each of the three studies the level of ambiguity aversion declined when participants were not given a chance to compare the two different bets. Study four followed the same logic as the first three, but had participants bet on a natural event rather than a game. Participants were asked to bet on the weather of either Istanbul or San Francisco. Finally, a study was conducted to see if individuals ambiguity aversion was affected by the suggestion that experts were being asked the same question.

In all the studies participants bet less money on the uncertain probability, demonstrating ambiguity aversion. However, when respondents were not given an opportunity to compare probabilities the effect disappeared, demonstrating that there is a comparative ignorance heuristic that influences individuals' levels of ambiguity aversion. In the comparative assessment of probabilities of natural events, participants placed higher bets on San Francisco, and in the non-comparative assessment the bets were more similar. Study four demonstrates that comparative ambiguity aversion can be elicited when probabilities are attached to a geographic area in which people are more or less familiar. The study also argues that the confidence with which individuals make risk assessments diminished when they compare themselves to "experts."

It is well known in the decision sciences that the human decision-making in the face of uncertainty relies on a decision-maker's interpretation of likely potential outcomes, and his/her beliefs regarding the probabilities of such events. The notion that outcome probabilities are merely subjective interpretations of uncertainty injects into the decision frame a material importance regarding the individual's recognition and understanding of the decision content. Further, the issue of subjectivity can greatly complicate matters when it comes to more complex problems involving multiple stakeholders. Building on Rittel and Webber's (1984) concept of wicked problems, Koppenjan and Klijn (2004) recognize substantive uncertainty as the difficulty of defining the nature of the public problems, which arises from actors' differing perceptions and interpretations of information. They present substantive uncertainty as one of three major sources of conflict in problem solving. Next, I briefly review the role of cognitive heuristics as a factor in human understanding of uncertainty, with a particular focus on the representativeness heuristic as presented by the classical studies of Kahneman and Tversky (1973) and Tversky and Kahneman (1983).

Representativeness heuristic

In normative statistical procedures of scientific prediction, such as climate forecasts, the following three types of information should be considered: (1) prior or background information (such as base rates); (2) specific evidence from an individual case; and (3) the analyst's expectation of accuracy in prediction. According to the normative theory, a predictor should follow a strategy where the relative weight of the first two types of information are based on the third: Higher expected accuracy should denote greater value placed on specific evidence, while lower accuracy should encourage greater weight on prior information. Representativeness is a general heuristic by which a predictor's intuitive judgment allocates weight between prior information and specific evidence differently than is prescribed by normative statistical theory. By such method, an outcome is judged to be more likely if it is more greatly representative of the specific evidence. This principle, like other heuristics, often results in adequate prediction, but may falter in certain cases, such as where specific evidence may be representative of an extreme or unusual outcome (Kahneman and Tversky 1973; Tversky and Kahneman 1983).

In their seminal paper, *On the Psychology of Prediction*, Kahneman and Tversky (1973) reported a series of studies focusing on either categorical prediction or on numerical prediction. In the categorical study, they sought to examine whether the subjects' intuitive predictions would violate the rules of normative statistical methods.

They hypothesized that where subjects should follow a Bayesian-type revision procedure (predicting likelihood beginning with prior information and updating according to specific evidence), they would instead base their predictions to a great extent on the specific evidence presented, sometimes ignoring prior information and expected accuracy entirely. The research question they posed to get at this hypothesis was whether or not the order of perceived likely outcomes would coincide with the order of representativeness of these outcomes with the evidence presented.

The study respondents were split into three groups, one establishing a base-rate, one establishing the similarity, and one predicting outcomes. The base-rate group was asked to estimate the percentage of all U.S. first-year students enrolled in each of nine fields. The similarity group was given a brief description of a person and asked to rank the same nine areas of specialization according to the similarity of the person in the description. The third group was given the same personality sketch and told that it was collected while the person was in high school, then asked to rank the likelihood that he were now a graduate student in each of the nine areas. Respondents in the third group were also asked to rate the percentage of the time they thought it would be possible to predict the correct field as the first choice (Kahneman and Tversky 1973).

Kahneman and Tversky (1973) found that there was a high correlation between the similarity rankings and the likelihood rankings, but that no relation was exhibited between the likelihood ratings and the base rate rankings. The subjects did not rely on their prior inclinations regarding the distribution of students among the nine disciplines, but rather made their predictions based very heavily on how representative the description was of each discipline. As previously discussed, statistical theory says that one can ignore base rate information when one expects his/her predictive accuracy to be absolute. However, the third group responded that they believed it possible to predict correctly about 23 percent of the time. Thus, even though respondents believed their predictive accuracy to be low, they systematically ignored the base rate information (Kahneman and Tversky 1973).

A second categorical study was carried out in which respondents were asked to rate descriptions of people on the likelihood of each to be a description of an engineer or a lawyer. Eighty-five subjects were split into two groups, who were given opposing base-rate data—half were told that the descriptions were drawn from a sample of thirty engineers and seventy lawyers, while the other half were told that the descriptions were drawn from a sample of seventy engineers and thirty lawyers. Half of each group were asked to rate the likelihood of each description being of a lawyer, and the other half of each being an engineer. The descriptions were each crafted to be characteristic of either an engineer or of a lawyer, with one of the descriptions being a null case, which provided no information indicative of either stereotype (Kaheman and Tversky 1973).

The authors found that when no evidence was given (rating the null case), the base rate information was utilized by respondents, but that when the specific evidence favored either profession, no matter how contradictory it was to the base rate, respondents ignored the base rate information (Kahneman and Tversky 1973).

In the numerical study, Kahneman and Tversky questioned whether intuitive judgment would follow the statistical rule that when predictors were more uncertain of their accuracy, they should tend to regress towards the base rate information. They sought to explore this by specifying the difference between evaluation of input data and prediction of outcomes—since evaluation is a task containing less uncertainty than the task of prediction, expected accuracy should be lesser in the predictive task, and such predictions should regress towards the base rate information. The subjects were provided with descriptions of college freshmen from their entrance interviews, and split into a group of evaluators—asked to assess their impressions of the descriptions—and a group of predictors—asked to predict future performance of the students. In the first study, subjects were presented with only five adjectives describing each student. In the second, they were given paragraph-length descriptions including background academic data and information about the student's adjustment to the college lifestyle. The authors observed that despite the reduced expectations of accuracy in prediction, the distribution of within-group output data presented similar standard deviations. Thus the evaluation and prediction groups produced equally extreme intuitive judgments, supporting Kahneman and Tversky's representativeness hypothesis (Kahneman and Tversky 1973).

A major limitation in Kahneman and Tversky's (1973) paper is their position that a judgment procedure, especially an intuitive one, should adhere to the rules of statistical inference. This normative assumption is a driving factor in their hypotheses; however, it discounts the element of subjective perception in subjects' creation of mental models of problems. Particularly in the light of more recent developments concerning the processes of two-system reasoning (Sloman 1996), it is fallacious to discuss deviation from the logic of one system (the rule-based, deterministic system) as "error." A focus on the failure of one system to control the entire process of intuitive judgment ignores a broader question of the relative influence of each system, which may differ among subjects depending on perception of the problem. Further, if the logical processes are taken as the normative ideal, the experiment should be measured across varying levels of expertise, and controlled for effects of programming (exposure to, and integration of, such statistical procedures) versus hardwiring. It should also be noted that a Bayesian type revision problem may not be appropriate in the case of predicting single-event probabilities, and as such, an intuitive process may recognize the problem as different from the causal learning, posterior-probability problem for which a Bayesian-type procedure would be appropriate.

A decade later, Tversky and Kahneman (1983) further explored representativeness, questioning whether the heuristic would override other rationally normative rules in intuitive judgment. They looked at the conjunction rule, which states that the intersection subset A and B, since it is included within both set A and set B, cannot be larger than either A or B. The authors sought to determine whether inclusion of an extra attribute (relegation to the intersection subset) added to a description would increase subjects judgments of correspondence between the descriptions and a list of occupations and avocations about the individual. Specifically, they wished to explore whether subjects would rate the conjunction "Bill is an accountant who plays jazz for a hobby" more likely than either of its constituents "Bill is an accountant" or "Bill plays jazz for a hobby."

In the study, eighty-eight undergraduate subjects were asked to rank statements of classes based on the likelihood that the description resembles a member of each class. The subjects were split into three groups. The main study carried out two types of tests:

> Subjects first performed a direct test, in which they were asked to rank several statements including both the conjunction and the conjunction's constituents; after which they performed an indirect test, in which they ranked statements in a similar problem, but this problem contained only the conjunction or its constituents, not both. The authors predicted that subjects would rank bill's likelihood of being an accountant highest, followed by his likelihood of being an accountant and a jazz musician, followed by his likelihood of being a jazz musician (A > A & B > B). They also predicted that subjects would recognize repetition of certain elements in the direct test, and apply the appropriate logical rules. However, Tversky and Kahneman found that in all comparisons, the subjects ranked the conjunction higher than its

constituents, and that there was no significant difference between the direct and indirect tests. The study supports the authors' hypothesis that intuitive judgments follow general heuristics which violate normative rules of statistical prediction.

(Tversky and Kahneman 1983)

In 1988, Gigerenzer, Hell, and Blank further elaborated on the dynamics of representativeness heuristic. The authors specify a representativeness heuristic (neglecting to use base rates) as opposite from a strategy of conservatism (using only the base rate, and neglecting specific evidence), and further argue that a subject's use of base rates is rather determined by an internal mental model representing the problem, and thus not limited to either extreme approach. They posit that a Bayesian-type updating strategy is not appropriate for many everyday, real-life problem-solving instances, and they further hypothesize such a procedure can be induced based on effects of problem presentation and content (Gigerenzer *et al.* 1988).

In the first experiment, the authors recreated Kahneman and Tversky's (1973) engineer-lawyer problem, with a modification in which subjects physically witnessed the random sampling of descriptions (by drawing them from urns before responding to the appropriate prediction problem). The authors wished to examine whether witnessing randomness would induce a Bayesian-type mental model of the problem, or whether the same base rate neglect effects would be seen that were present in the original study. Ninety-seven students were split into two groups, one of which visually observed the random selection of the descriptions presented, and the other group was given only a verbal assertion of randomness as in the original study. The authors found that for the group that visually observed the random sampling, the deviation from a representativeness strategy increased from the verbal assertion group, and that deviation from a Bayesian-type prediction decreased with observance of random sampling (Gigerenzer *et al.* 1988).

Further, in the experiment, the subjects were asked to describe their strategies for arriving at their predictions and to describe how they chose to utilize the base rate information regarding the proportions of engineers and lawyers. The strategies were classified by two observers into five general categories: base rates only, Bayesian revision, Bayesian integration (where both types of information were considered, but not via revision), lexicographic representativeness (where similarity was considered first, and base rates were consulted only if necessary), and representativeness heuristic (where subjects followed the classic base rate neglect strategy). The choice of strategy was different depending on whether the subjects were in the verbal assertion or visual observation group. More than twice the percentage of respondents in the visual group reported following a base rate only strategy than in the verbal group. Similar percentages of respondents in either group chose the Bayesian revision and lexicographic strategies.

However, a significantly higher portion of the visual group chose a Bayesian integration strategy, and a significantly higher portion of the verbal group chose a representativeness heuristic strategy. These results support the difference that was observed in the predictions, that the visual group tended towards a Bayesian strategy, and the verbal group tended towards a representativeness strategy, further supporting the hypothesis that the presentation of the problem influences the subjects internal working model of the problem and the appropriate strategy of prediction (Gigerenzer *et al.* 1988).

The second experiment sought to explore the effects of content on the internal representation of prediction problems. The authors again attempted to disprove Kahneman and Tversky's (1973) conclusion that the human mind is ruled by a general representativeness heuristic by using a problem based on content more familiar to everyday life in order to induce a Bayesian-type revision procedure, as prescribed by normative rationality under uncertainty (statistical theory). The experiment was designed so that subjects were given data on soccer teams, and asked to predict a game's outcome once at the beginning of the game, and then again at half-time. The subjects were given information on the number of games previously won by the participating team, and on the percentage of times the team was leading at half-time and subsequently won. The subjects were split into two groups, one of which was given varying base rate and varying specific evidence, while the second was given varying specific evidence, but consistent base rate information. As in the first experiment, they were asked to report their strategies for predicting the outcome of the games (Gigerenzer *et al.* 1988). The authors found that in comparing the soccer problem against the engineer/lawyer problem, a greater portion of the subjects followed both types of Bayesian strategy, while a significantly lower number followed the lexicographic strategy, and zero percent followed the representativeness heuristic. In the engineer/lawyer problem, the strategies were more heavily weighted towards the representativeness heuristic. The results support the authors' hypothesis that *subjects' familiarity with the content of a problem does influence the choice of strategy, and that increased familiarity will increase the use of a Bayesian-type procedure in predicting outcomes.* In general, the results exhibit the presence of a representativeness heuristic, but do not prescribe it as a rule for intuitive judgments. Instead, the authors argue, such a heuristic is one of several strategies along a continuum, the choice of which depends on other factors (framing and content of the problem), which help the mind recognize a problem as appropriate for a Bayesian updating procedure versus a conservative or representativeness strategy (Gigerenzer *et al.* 1988).

In a recent study, Ledgerwood *et al.* (2010) expand the question of representativeness and problem perception to include the effects of construal level. The authors define construal level to be the representation of psychological distance (temporal or geographic) in the construction of a mental model of a problem. The study sought to explore whether the construal level would affect how subjects implement available information (i.e., how they choose to weight prior information versus specific evidence) in prediction problems. The authors hypothesized that increasing psychological distance would increase reliance on aggregate statistical information (e.g., base rates) whereas decreasing distance would shift weight towards context-specific evidence (Ledgerwood *et al.* 2010).

Numerical probabilities are commonly used in weather and climate forecasting, although an earlier study by Murphy *et al.* (1980) found that the majority of people surveyed misunderstood numerical probabilities used in weather (and now climate) forecasting. This misunderstanding was directly attributed to a failure to identify the reference event of the probability. Extending the findings of Murphy *et al.*'s research, Gigerenzer *et al.* (2005) hypothesized that a person's ability to understand weather forecast probabilities is a result of an individual's exposure and length of exposure to these types of forecasts. To test their hypothesis, Gigerenzer *et al.* (2005) conducted a survey in five major cities in the United States and Europe (Amsterdam, Athens, Berlin, Milan, and New York City) that use numerical probabilities in weather forecasts in varying capacities. Survey participants were asked what "a 30 percent chance of rain tomorrow" meant and at what percentage they would carry an umbrella with them around town. It was found that individual exposure to weather forecast probabilities had little to do with their ability to understand them and most people would carry an umbrella if the probability of rain were forecasted at 50 percent or higher. Within the cities where the survey was conducted, weather probabilities were presented in a variety of ways (e.g., verbal only, numeric only, a combination of verbal and numeric). The use of a particular probability expression was not always consistent over time. This study specifically looked at how individuals interpret a seemingly straightforward numerical probability for a single event. The participants were not provided with any contextual information when they were asked the question. Gigerenzer *et al.* (2005) concluded that numeric probabilities should be accompanied by clarifying text that specifies the exact reference class of the event to improve individual understanding of the probability.

Fifteen years earlier, Weber and Hilton (1990) came to the same conclusion when they recommended using a combination of verbal and numerical probability expressions to limit the likelihood of miscommunication amongst people. Their research was stimulated by the assertion that the interpretation of probabilities is contextual and is influenced by how likely an individual thinks the event is to happen, otherwise referred to as the base rate probability. Building upon previous experiments conducted by Wallsten, Fillenbaum, and Cox (1972) and Svenson (1975), Weber and Hilton (1990) evaluated the difference in interpretation of verbal uncertainty expressions as a result of perceived base-rate probability, as well as the perceived severity of the event. The experiment was conducted with undergraduate students at the University of Illinois. Participants were presented with a series of statements that combined a medical condition and verbal probability expression and were asked to determine the numerical probability that a doctor may have in mind when they made that statement. Afterwards, they were asked to rate the likelihood that they thought they would actually develop each of the medical conditions in the next year and how serious they perceived each condition to be.

In a follow-up experiment, the exact same methods were followed except the participants were asked to rate how likely it was for someone similar to them to develop the medical condition in the next year. This question was specifically aimed at evaluating whether an individual's base rate probability of an event changed when the situation was no longer personal in nature. As expected, the base rate probability increased in the second experiment, confirming that people tend to think an event is more likely to happen to someone else than to themselves. In both experiments, it was found that the perceived base rate probability and severity of the condition affected the numerical value assigned to each statement, but it was not possible to ascertain the extent to which each variable affected the outcome. Weber and Hilton (1990) recognized that assessing too many variables at once complicates one's ability to assess which factor is influencing the numerical evaluation of a verbal expression and to what extent.

The perception of risk, or the expectation that an event will happen, was found to be a motivating factor for water managers to use weather and climate forecasts. O'Connor *et al.* (2005) surveyed water managers in South Carolina and the Susquehanna River Basin of Pennsylvania during the summer of 2000 to ascertain why water managers did or did not use forecasts to inform their decision-making. The researchers believed that the decision to use forecasts depended on whether there was a perceived risk and whether the manager was confident in the accuracy of the forecast. The survey asked water managers how weather and climate forecasts were used to plan and budget their projects, how reliable the manager felt that forecasts were, whether the managers expected to experience certain severe weather events in the next ten years, and whether they had experienced the same severe weather events in the past five years. Although the perceived reliability of the forecast did not affect whether forecasts were used, the perception of risk was found to be a driving factor for water managers to use forecasts; further, this perception of risk was often tied to previous experiences with severe weather events.

Past experience may increase the perception of risk associated with a particular event when an individual can remember details about a similar event in the past, making the possibility of a similar event happening in the future seem more plausible. According to Leiserowitz (2006), risk perception and behavior is guided by personal affect (e.g., emotions and connotations), imagery (i.e., visual cues), and worldviews. As discussed earlier, Leiserowitz (2006) found that Americans do not perceive climate change as a local issue, rather Americans are primarily concerned with the effect of global warming on other people in the world and on nature in general. According to Leiserowitz (2006), "most Americans lacked vivid, concrete, and personally-relevant affective images of climate change, which helps explain why climate change remains a relatively low priority" for the United States as a whole (Leiserowitz 2006: 55). Without a clear personal tie to the potential consequences of climate change, Americans were willing to support policies at the national and/or international level, but were not keen on backing more local policies that would directly affect their livelihood.

Availability heuristic

The availability heuristic refers to how readily an individual can bring an image or association of the risk to mind, reflecting how salient a particular issue is to them. Climate change is often thought of as a future problem with the expectation that significant consequences may not be realized for several years. Sunstein (2006) questioned why, if this were the case, Europeans were strong advocates for taking immediate action, while Americans were not actively trying to address climate change. To answer this question, Sunstein (2006) discussed how individuals use a form of cost-benefit analysis, in addition to the availability heuristic, to assess whether the cost of taking action is worthwhile with respect to the benefit they will see. More specifically, the cost of addressing climate change is particularly high for the United States and the perceived benefit of reduced greenhouse gas emissions is relatively low, compared to other nations. Drawing heavily on the idea of the availability heuristic, Sunstein emphasized that it is difficult to actually imagine what will happen as a result of climate change and older generations may never experience the consequences. He cautioned, however, that, "the availability heuristic can lead to serious errors, in terms of both excessive fear and neglect" (Sunstein 2006: 198).

According to Weber and Hilton (1990), the severity of an event can also be thought of in terms of utility; if the experience of event is expected to be positive, the perceived likelihood increases and vice versa. In this regard, individuals likely underestimate the potential consequences of climate change because most of the consequences are portrayed in a negative light. The use of numerical percentages in communicating risk may also contribute to a misunderstanding depending on the quality of the contextual information provided and whether a clear reference event is defined. Without a tangible event determined to be the result of climate change, the American public lacks an experience by which they can form strong images in their minds or emotions in their hearts to solidify the reality of the risk of climate change. Until a strong event association is readily available for individual Americans to grasp the breadth of the impact of climate change, individuals may continue to feel that it is more likely that climate change will affect someone else, and as such may be slow to support policies specifically created to mitigate the causes of climate change right here in the United States.

Framing effects

A critical assumption in rational models of behavior is that decision-makers will make similar decisions, regardless of how the decision problem is presented. This assumption is called the principle of invariance. In the early 1980s research began to emerge that showed that the principle of invariance does not hold true in risky decisions. In risky decision-making, people's preferences change based on how decision problems are framed and presented.

A key insight that we can take away from framing is that information is not objective. Each piece of information that a decision-maker receives is only a small slice of the context that a decision-maker needs to make a "rational" decision. The way these pieces of information are fit together and communicated can produce illusions that promote one option over another. These illusions are what Tversky and Kahneman (1986) call framing effects.

This widely cited article by Tversky and Kahneman (1986) was a response to rational actor models of behavior. Tversky and Kahneman (1986) thesis is that the "logic of choice" is not adequate as a foundation of theories of decision-making under risk. They carried out experimental research that was designed to test the principle of invariance, which is one of the assumptions in rational models. The principle of invariance (in normative decision-making models) claims that different representations of the same problem will yield the same preferences. They carried out a series of twelve experiments on students at two universities. Each of the experiments contained two or more options, and one or more of these options had uncertain outcomes. Furthermore, there were two groups of options with the decision problem framed differently (for example, as a gain or loss). They define framing as "the manner in which the choice problem is presented as well as by norms, habits and expectancies of the decision-maker" (Tversky and Kahneman 1986: 8257).

Tversky and Kahneman (1986) found that decision problems are not considered in all angles of light and transformed into a "common frame" when decisions are made. Tversky and Kahneman (1986: 8256) conclude that "variations in the framing of decision problems produce systematic violations of invariance and dominance that cannot be defended on normative grounds." They liken framing effects to visual illusions, rather than a calculation error by the decision-maker. They found several different effects that framing has on decision behavior. In the first decision problem they present, they frame the decision problem as a relative loss or a relative gain. They found that choices involving losses are usually risk seeking and choices involving gains are usually risk averse. Based on these findings they suggest a value function that is S shaped. It is defined by gains and losses, it is concave for gains and convex for losses and steeper for losses than for gains. The certainty effect and the pseudocertainty effect are the phenomenon that when there is certainty in one option decision-makers tend to be risk averse, and when there is significant uncertainty decision-makers tend to be more risk taking. Tversky and Kahneman (1986) found that in two-stage uncertain decision problems people tend to disregard the first stage and base their decision on the second stage. This leads to a violation in the principle of invariance, which they call the cancellation effect.

In a meta-analysis to assess the influence of framing on risky decisions, Kühberger (1998) found a "small to moderate" size framing effect present in the overall body of research on framing effects. Kühberger (1998) identified nine distinct research designs used to elicit framing effects. Risk can be differentiated in three ways in experimental design: (1) risk manipulation; (2) quality of risk; and (3) number of risky events. Overall, the most important characteristic that produced framing effects in research was that of risk manipulation. Kühberger (1998: 44) concludes that "framing research may have a lot to say about more practical aspects of judgment and decision in daily life."

Durfee (2006) studied the role of the media in communicating risks associated with unhealthy air quality in Salt Lake City. Specifically, Durfee (2006) looked at how media frames of risk regarding air quality influenced public perceptions on risk. Durfee (2006) studied two "types" of frames: "status quo" frames and "social change" frames. Status quo frames, in this case, discuss air quality issues as a weather event, that has no bearing on people's behavior. The social change frames attribute responsibility for air quality to individuals and their travel behavior. Durfee (2006) surveyed 140 undergraduate communication students at the University of Utah. Each survey contained a news article displaying a social change frame or a status quo frame. Durfee (2006) found that participants that read the story with a social change frame had the highest level of risk awareness.

In a recent study by Spence and Pidgeon (2010), 161 psychology students were surveyed and randomly given either a gain frame or loss frame and a local or distant frame about the risks of climate change. Gain frames produced more severe judgments of impact than loss frames, and produced more positive attitudes towards climate change mitigation. They found that participants who received the gain frame were more receptive of the information compared to the participants that received the loss frame.

The distant frame elicited judgments of higher severity than the local frame; however, this did not lead to any difference in attitudes towards mitigation. The local frame had higher ratings of personal relevancy than the distant frame. Spence and Pidgeon (2010) conclude that "communications promoting climate change mitigation should focus on what can be gained by mitigation efforts rather than dwelling on the potential negative impacts of not taking action."

A curious discovery of modern behavioral economics and decision theory is that many so-called rational actors prefer to minimize an expected loss rather than maximize an expected gain, and that such actors also prefer a solution that minimizes *both* loss and gain to one that positions probabilistic maximal gain against similarly uncertain larger loss. In its traditional form (i.e., rational economic actor), loss aversion assumes an unreflective experience effect, where experience with loss aversive behavior—either positive or negative—has no influence on future decisions regarding loss; or, that loss aversive behavior is compulsive and non-introspective. Put simply, the decision heuristic of loss aversion is one that "expect[s] outcomes to be coded as gains or losses relative to a neutral reference point, and [for] losses to loom larger than gains" (Kahneman and Tversky 1979).

In literature relating to loss aversion, there is a concerted effort on the part of the majority of researchers to distinguish economic loss aversion as a particularly aberrant, irrational behavior, one with damning consequences for the rational economic actor. Loss aversion casts the understanding that a given actor maximizes his/her expected utility in an unfavorable light by suggesting that maximized utilities in practical application are not only not objectively derived mathematical quantities wherein the most optimal outcome within given trade-offs is sought, but also that the trade-off costs—rather than optimal outcomes—are the driving force behind decisions exhibiting paradigmatic loss aversion.

Of primary concern herein is the establishment that pain and loss aversion are similar or nearly identical behaviors, specifically for the purpose of generalizing pain aversive behavior as it applies to decisions under questions of uncertainty. As such, it is critical to draw together the major dichotomy in loss aversion: that economic loss aversion is irrational in a unique way as compared to pain aversion in other contexts.

The first consistent theme that emerged from Loewenstein's 1996 investigation draws a comparison between loss aversion and pain aversive behavior; namely that they are comparable behavioral drives with commensurate applications. The thrust of this theme is that in decisions that are susceptible to "hot" or "cold" states of emotion—what George Loewenstein refers to as "visceral influences"—the ability of a decision-maker to account for his/her long-term optimal utility is obscured by a near-term "urge" or "visceral factor" that overrides long-term planning in lieu of satiation. These "visceral factors" are broadly understood to have "first, a direct hedonic impact (which is usually negative), and second an influence on the desirability of different goods and actions. Hunger, for example, is an aversive sensation that affects the desirability of eating" (Loewenstein 1996).

Dan Ariely (2009) points out that social science experiments are best regarded as "illustration[s] of general principle[s], providing insight into how we think and how we make decisions," rather than as hard-and-fast data generation machines. Though Loewenstein's 1996 article is limited on hard data, it is nonetheless an extremely valuable resource for addressing the central heuristic at issue here, which is that pain aversion and loss aversion are essentially the same or at the very least directly comparable.

Jenni and Loewenstein (1997) attempted to test what is known as the "identifiable victim effect," which is that "society is willing to spend far more money to save the lives of identifiable victims than to save statistical victims." In this case, the loss aversion is characterized by the desire to avoid the loss of a known victim or victims as it is contrasted with the loss of unknown victims; essentially, certain loss versus uncertain loss. Jenni and Loewenstein (1997) conducted two studies testing four possible explanations: vividness, certainty and uncertainty, proportion of the reference group saved, and *ex post* versus *ex ante* evaluation. They found statistical correlations for certainty and uncertainty as well as the proportion of the reference group saved, but both vividness and *ex post/ex ante* evaluation failed to provide compelling evidence for their influence over the identifiable victim effect.

Harinck *et al.* (2007) investigate the scale effect of the loss aversion paradigm, and in particular treat "loss aversion as an *affective forecasting error*" [emphasis in original], which can "not only be diminished, but may actually be reversed" so that "small gains loom larger than small losses." The central argument of Harinck *et al.* is that loss aversion as a decision heuristic does not operate in a vacuum relative to other influences, and is particularly susceptible to scale effects such that small losses are more common than large losses, so people ... are aware that they can overcome small losses and that their actual reaction to such losses is not as negative as they imagine beforehand. People may have greater difficulty anticipating their true reactions to large losses, however. ... Therefore, the impact bias is more likely to be present for large negative outcomes than small negative outcomes: people anticipating a large loss will tend to overestimate the negative feelings that they will experience

(Harinck *et al.* 2007: 1100)

Another major aspect in attempting to unravel the tangled web around loss aversion is the tendency for agencies or organizations that distribute information intended to inform agents regarding their decisions to engender the use of technical jargon. Though jargon may be more specific than plain English for the purposes of addressing specific aspects of a given decision problem, the density of such jargon can obscure loss aversive behavior by either pushing agents to err on the side of overly conservative decision-making (or to refuse to decide altogether) or to make a decision inconsistent with their intentions. The studies by Fagerlin *et al.* (2005) and Golding *et al.* (1992) on informed consent are indicative of the above quandary.

Fagerlin *et al.* (2005) set out to compare several methods of delivering informed consent information to angina sufferers regarding two potential treatments, and ways to combat situations in which "the patient [rejects] statistical reasoning in favor of anecdotal reasoning" (Fagerlin *et al.* 2005). The study compared presentation of information in two parts: standard prose statistics versus "enhanced" statistics, the latter containing a graph and quiz while the former stated the data as plain text; and anecdotes/testimonials in which the ratio of positive experiences to negative experiences was either representative or unrepresentative of surgical success rates. The primary discovery is that

enhancing standard prose statistics with a pictograph can reduce the influence of anecdotal information on people's hypothetical treatment choices. That is, participants who received a pictograph illustrating the statistical information about each treatment were not differentially influenced by representative versus nonrepresentative anecdotes.

(Fagerlin *et al.* 2005: 404)

This implies that the loss aversive behavior can be elicited through the use of differing information delivery techniques. This result is not in itself significant to the heuristic of loss aversion, but compare the obfuscating influence of anecdotal evidence when it is held constant, in which of "those who received standard prose statistics (no pictograph and no tradeoff quiz), 27 percent chose bypass surgery, whereas 37 percent chose bypass surgery among participants who received statistical information enhanced by both a pictograph and a tradeoff quiz" (Fagerlin *et al.* 2005: 404).

Bearing in mind that the trade-offs in the hypothetical scenarios for the two treatments are that bypass surgery is successful 75 percent of the time and requires a week's stay in the hospital, whereas balloon angioplasty is successful 50 percent of the time and requires only an overnight stay, a clear distinction emerges: in the face of a significant life decision, the word-of-mouth heuristic powerfully influences loss aversive behavior; but, when the word-of-mouth heuristic is removed, the majority of respondents favored the method of treatment that *minimized their near-term loss, despite the potential long-term gain.*

The focus of the Golding *et al.* (1992) study is to measure the utility and quality of the information delivered on the dangers of radon by two methods of communication published in local newspapers: the first, a narrative in several parts detailing a mother-of-two's investigation of and discoveries regarding the dangers of radon in her home; and the second, a technical account of the dangers of radon taken more or less intact from government documents. The goal of the study is to assess which of the two methods, given that they were read, is more effective in disseminating necessary information to the general public. The effectiveness of this dissemination is measured by focus group sessions and telephone canvassing.

Though not directly applicable to loss aversion, the findings in Golding *et al.* (1992) indicate the pervasiveness of the phenomenon and introduce an interesting tension with regard to loss aversion that has not seen enough treatment from a research standpoint: the attritional apathy generated by continuous loss aversion behavior. This apathy is best encapsulated in so-called "worry budgets," in which homeowners prioritize possible radon infiltration as "just another new worry in a long and growing list of environmental and lifestyle hazards" (Golding *et al.* 1992: 32). In other words, because radon is a specific problem with known characteristics, easily tested for and mitigated, radon infiltration itself does not approach the loss aversion threshold; but the "severe distrust of private testing and mitigation companies and the fear of being misinformed or even hoodwinked" (Golding *et al.* 1992: 32) *does* satisfy the average respondents' criteria for a loss aversive behavior. The aforementioned attritional apathy dictates, therefore, that a concerned homeowner avoid the near-term financial loss that may be incurred by a fraudulent radon testing/mitigation company in lieu of testing and mitigating for radon, with the understanding that the near-term financial loss would be unacceptable in the face of the relatively paltry gain of knowing one's radon infiltration levels (as few homes have a severe radon infiltration worthy of mitigating).

The two major tenets of this review are that loss aversion, as a decision heuristic, is both insidiously visceral—loss aversion and pain aversion are commensurate measures—and easily obscured by other heuristic influences or miscommunication. At baseline, however, loss aversive behavior exhibits these three major characteristics:

1   Unconscious or subconscious action, such that loss aversion comprises the root motivation for many actions.

2   Baseline preferential superiority, though it may be obscured by verbal recontextualization or rational level heuristic influence.
3   Threshold specificity, whereby small losses versus small gains are not considered significant enough to be subject to idiosyncratic loss aversion.

It is in this context that I argue that emphasis on climate change induced losses can trigger a trans-ideological mitigation and adaptation response in the long run. Despite the recent evidence supporting the occurrence of heat waves, droughts and wildfires in the United States and EU that could be attributed to anthropogenic climate change in the last two decades (e.g., Hansen *et al.* 2012), the larger media discourse on climate change has dampened since the onset of 2008 global recession. Perhaps the larger loss frame of global recession has dominated the other loss frame of climate change induced droughts, floods and heat waves. Yet, there is a connection between the two as well, which largely goes missing in the public discourses.

Reframing climate science communications

Zia and Todd (2010) show that many conservative ideologues are less concerned about climate change science because they believe that scientists do not have adequate/clear understanding of climate change science. The strong significance of the variable "science unclear" in Table 4.2 for global warming concern testifies to this problem. Further, Figure 4.3 shows a very strong correlation (31 percent) between "science unclear" and "ideology," i.e., as ideology shifts from liberal to conservative, survey respondents are more likely to believe that climate change scientists are not clear about understanding the climate change science. This belief among the conservative ideologues has to do with the longstanding discourse in the United States about whether climate change is human-made (anthropogenic) or "natural." Many conservative ideologues tend to believe that climate change is natural and not anthropogenic, thus society and governments need not take any action to deal with climate change. While systematic efforts have been put in place by climate change scientists, especially after IPCC was awarded 2007 Nobel Price, to dispel the belief that climate change is merely "natural," conservative ideologues (as shown in Zia and Todd's (2010) spring 2008 survey) tend to disbelieve the IPCC consensus that climate change is anthropogenically induced through the momentous release of greenhouse gas emissions in the atmosphere since the beginning of industrial era.

Previous researchers have noted that messages about climate change "need to be tailored to the needs and predispositions of particular audiences; in some cases to directly challenge fundamental misconceptions, in others to resonate with strongly held values" (Leiserowitz 2006: 64). In this context, Zia and Todd (2010) recommended that scientific forecast messages must be further tailored to cut across ideological predispositions. Moser and Dilling (2004), for example, urge public communicators to address the emotional and temporal components of "urgency," and tap into individual and cultural values, in order to conquer ideological differences. Specifically, Zia and Todd's (2010) findings demonstrate that conservative political and religious ideologues are least likely to name climate change as an important present or future concern. Returning to the "guns versus butter" theory, Zia and Todd (2010) identified three strategies of climate change communication that might be used to more specifically appeal across the conservative–liberal ideological divide. They discussed how security appeals can potentially frame climate change as a "guns" issue. They then explored how religious appeals can frame climate change as a "butter" issue. Finally they suggested how framing climate change mitigation and adaptation as part of a human response to the economic crisis might cut across ideological lines.

A significant trend in public discourse is framing climate change as a security issue. This is evident in President Obama's efforts to add experts on "climate security" in the membership of the National Security Council. "Energy security" is a loaded phrase used by oil companies and grassroots groups alike. Ungar (2007: 85) notes that links between security concerns and energy supplies in public discourse is a potentially effective strategy at gaining public adherence. Climate change can be framed as more of a security issue by emphasizing the risks and impacts of drastic climatic change. Climate change is an invisible threat, with no immediate proof of its danger, or existence for that matter. Since "security" is a guns issue, ideologically predisposed conservatives might have higher concern about "climate security" as opposed to "climate change." As a guns issue, climate change is positioned as more relevant to citizens' personal security. Appeals to security are persuasive when threats seem real. In the same way that the color-coded terror alert levels explain the need for security procedures, climate security threats could be publicized to justify government action. This could be accomplished by incorporating evidence of tangible effects of anthropogenic climate change into discussions of present and future threats to environmental security. This would link present policy to concerns of future impacts, and reframe climate change as a direct threat to security.

A second trend in public discourse about climate change is reframing climate change as a "religious" issue. James Dobson and Reverend Sally Bingham are among American religious leaders calling for clergy to advocate more action on climate change (e.g., Bingham 2007: 153). Since conservative ideologues are typically "religious minded" (as is obvious from cultural models elicited in Figure 4.3), it might help to improve the communication of climate change science across ideological divide if climate change is framed as an issue of "pain and suffering for fellow humans, animals and plants." As a moral issue, climate change is important to religious communities (see Leiserowitz 2007). As in the case of "climate security" reframing, theologizing of climate change scientific communication might have some appeal for conservative ideologues. Climate change can be framed as a "butter" issue through appeals to religious values, such as a moral obligation to protect the earth and God's living things, and alleviating suffering. Mitigation and adaptation strategies may be framed as ways to protect communities and as part of human moral obligation to future generations. As a moral issue, climate change becomes more of a question of compassion.

Such communication strategies could bring climate justice issues more to the forefront of the public mind and provide a more global perspective on the impacts of climate change.

A third strategy for improving communication of climate change along ideological lines is to explicitly frame climate change as an economic issue. The long-term economic effects of climate change were extensively documented in the widely read Stern Report (Stern 2008). Yet, for many citizens, these remain long-term effects with little relevance for today's consumers. In this way, economic issues are related to questions of certainty: the uncertainty of climate change impacts works to distance the economic impacts as well. Yet, while scientific uncertainty remains high for conservative ideologues, reframing of climate change across economic lines can tap into the emotional urgency of the economic crisis and create a new sense of certainty: the need for the certainty of action. Connecting climate change and economic recovery will effectively reframe the issue as a butter issue, while also appealing to conservative economic beliefs. The global economic crisis has extreme emotional and temporal urgency as citizens see their investments dwindle and the future become more uncertain. Connecting the climate crisis to the economic crisis and climate crisis frames mitigation as an investment in the future of society and adaptation as a protective measure for the well-being of citizens. For example, eliminating waste is an important aspect of economic recovery: witness President Obama's elimination of expenditures in the federal budget, and perhaps more poignantly, the state of California's warehouse sale. Framing energy efficiency as a cost-cutting and waste-saving measure is a strategy to persuade conservative ideologues of the importance of climate change. Questions about the costs of scientific investigation, and the costs of greenhouse gas mitigation can be framed as an investment in green technologies, or weighed against the future cost of predicted effects on agricultural industries. The economic crisis has altered America's (and the world's) confidence in short-term economic policies, and has increased receptiveness to long-term economic investments. Explicitly connecting our response to climate change to economic issues will tap into the personal and national importance citizens attach to the economic crisis. Framing climate change policy as a stabilizing response to economic uncertainty can mitigate questions about scientific uncertainty.

While careful research needs to be conducted to evaluate the effects of reframing the climate change discourse in security, religious, or economic terms, the re-framing of climate change discourse as a security or a economic issue might help to successfully communicate the climate change science across ideological divides. These strategies for improving communication about climate change respond to institutional and social changes such as those affecting international security and economic markets. Framing climate change as a problem related to these issues will help to affect behavioral change in individuals across ideological divides. Framing climate change policies as part of broader public discourse about international and economic security may increase support of more climate-neutral behavior in ways that are amenable to those with different ideological stances.

Zia and Todd (2010) demonstrated that ideology results in significantly different concern, and by implication understanding, of citizens for different policy issues that are at the intersection of science and politics. Ideology even trumps college education in the formulation of citizen cultural models. Zia and Todd (2010) demonstrated utility of mental models in assessing public understanding of climate change. Public knowledge of climate science that cuts across ideological divides is crucial to public action toward addressing these impacts. Reframing climate change knowledge along a guns versus butter political agenda theory, they argued, can potentially empower people across ideological divides to act locally on a global issue. However, the effects of reframed risk communication must be carefully investigated in future studies that also take into account cognitive dissonance and trust-gap effects.

Follow-up research must investigate whether and how changes in the informational content of scientific messages affect a change in the concern of citizens across a larger variety of socio-political and economic issues. Another question of interest is how long do citizens remain committed to their religious and/or political ideologies after scientific messages are tailored to unravel unscientific beliefs of citizens. Such research would be worthwhile for understanding the effects of ideology in communication of other kinds of science to the public, such as pesticide exposures; DDT effects; impacts of tobacco, alcohol and other drugs; biotechnology, nanotechnology and other cutting edge scientific discoveries. As society at large gains more scientific knowledge, we must find ways to disseminate that knowledge across the ideological divides to engage citizens in matters of public policy. This brings us to the politics of knowledge in communicating climate change risk as well as designing policy and governance regimes to cope with global climate change, which is discussed in Chapters 6 and 7.

6 The politics of knowledge I

Marketization of climate governance

Governance design: balancing government, market, and society relations

This chapter deploys the politics of knowledge lens to analyze the policy debates that surround the foundational policy and governance design questions concerning government, market and society relations for setting up effective and fair climate mitigation and adaptation policies. In particular, marketization of climate governance that was attempted in the Kyoto Protocol through the so-called flexibility mechanisms (CDM, JI and later on REDD+) is addressed from the perspective of politics of knowledge. A marketization approach to mitigate GHG emissions is one of the many available options on the table and it reflects a particular form of knowledge, in particular a neo-classical form of economic thinking, that might not be adequate in solving a climate change problem that has become a problem in the first place due to the "free market" activities in energy, transportation, agriculture, manufacturing and other GHG emitting sectors.

In the next section, a meta-theoretical discussion on the valuation of environmental goods and services, which includes the value of a "clean" atmosphere, is undertaken to demonstrate the conflicting knowledge claims that arise from favoring marketization based governance design approaches over a more balanced (or in other words "pluralistic") societal-market based governance designs. The marketization-based climate governance approach is intertwined with the perennial issue of fair GHG emission entitlements. On the one hand, developing countries contest that GHG per capita entitlements must be put in place. On the other hand, the developed countries disagree with this approach by asserting that their actions alone will not reduce the GHG emissions as it will merely shift the GHG polluting activities from the developed to the developing countries, especially given the larger population waiting to be "developed" in the twenty-first century.

In the following section, the politics of knowledge lens sheds light on this "wicked" climate policy GHG emission entitlement dispute and elucidates the politicized nature of climate science in an atmosphere of growing distrust and uncertainty. Further, in the final section, tropical forest governance is revisited from the perspective of one of the key drivers of tropical deforestation:

international trade and foreign direct investments, which are protected under the current "free market" driven trade-regimes of WTO and other Bretton Woods institutions. Balancing global trading and market interests against the local and national scale societal values in the tropical countries could be a key factor in re-designing post-Kyoto climate governance regimes that are not blindly enslaved to the "marketization" based politics of knowledge observed in the current REDD+ policy design.

Theoretical clash over environmental valuation

A theory of valuation typically addresses these types of questions: how the elements of a system in a decision-making problem are valued, to what extent, and by whom? How the values are prioritized and accounted for? What is not being valued (but could be)? Which values are in tension/conflict? How are the values traded off? Traditionally, valuation theories have been proposed along disciplinary lines. Theories of economic valuation such as Total Economic Value (TEV) or Capital Asset Pricing (CAP) postulate certain testable hypotheses that reduce the complexity of human values to measureable chunks of utilitarian calculus and observable monetary units (e.g., $s). In contrast, ethical theories of valuation, such as Dewey's Theory of Valuation (DTV) (1927), or a more recent "Deliberative, Pluralistic, Multi-Scalar" (DPM) theory of valuation, proposed by Norton and Noonan (2007), aim at understanding valuation processes from moral and socio-psychological perspectives.

I do not reject at the outset any specific disciplinary theories of valuation. Rather, I propose a broader, inclusive meta-theory of valuation that enables understanding of valuation from multiple disciplinary perspectives with the explicit precaution that each of these disciplinary perspectives provides a partial representation of what gets valued, by whom and how. The meta-theoretical perspective enables governance designers to be humble about the limitations of each of these valuation theories, while providing them with some powerful analytical tools to make trade-offs explicit in a decision-making process.

Valuation theories can be viewed from many dimensions, including (1) commensurability, (2) consequentialism, (3) scale, (4) process and (5) power, etc.

Commensurable Valuation Theories, such as TEV or CAP, assume that all values can be measured/transformed on a common scale (e.g., utility or price) for analytical tractability. Non-commensurable Valuation Theories, such as DPM, assume that values cannot be necessarily measured/transformed on a common scale, though they can be weakly compared.

Consequentialist valuation theories assume that the consequences of human actions are measurable, even in the face of risk and uncertainty. Non-consequentialist valuation theories, on the other hand, consider the assumption of consequence measurability as impractical on many grounds, starting from Ulrich's (1998) boundary critique theory to Simon's (1982) satificing theory about computational/cognitive limitations of human decision-makers. Most of the research in decision sciences has strived to unravel the underlying explanatory variables that cut through the "rational" strands of consequentialist valuation theories.

The treatment of scale also varies significantly across the spectrum of valuation theories. Typically, commensurable and consequentialist valuation theories allow for space-time discounting by human decision-makers in the system. Space-time discounting raises issues of boundary selection, choice of discount rates and ignorance about the value functions of those decision-makers that are not empirically verifiable/falisfiable (e.g., future generations, or animals), as discussed in Chapter 2. The intractability of these issues lends to a different treatment of scale in non-commensurable and non-consequentialist valuation theories, which emphasize the importance of *fair process* to transcend the questions of scale.

Without analyzing all of the tensions among different valuation theories, as this will require a separate book, here I emphasize one aspect of these tensions: cross-scale asymmetries. It is arguable, for example, that there is an inherent and unavoidable conflict, in many situations, between the goals of environmental conservation and economic development: When environmental conservation goals are global, as in attempts to protect worldwide biodiversity and carbon stocks in the tropical forests, the benefits of effective efforts in local situations often accrue globally, while many of the costs are felt locally. Zia *et al.* (2011) demonstrate this politics of scale for the case of Tanzania, but such politics of scale have been documented in many other tropical countries as well. One wicked challenge is that this politics of scale is intertwined with the politics of knowledge, whereby the dominance of one form of knowledge (e.g., neoclassical theory of economic valuation) dominates the governance and policy design considerations where environmental valuation needs to be undertaken, while the indigenous knowledge of hundreds of tribes living in the tropical countries who have actually co-existed with the tropical forests for centuries is undermined.

While environmental conservationists have been willing, from the days of John Muir and Gifford Pinchot, to argue strongly for the overall benefits of conservation over unwise development and destruction of natural systems, newer formulations of the controversy radically shift the *value* aspects of the debate. While arguing for protection as opposed to development in general, conservationists could engage successfully in comparisons of total values by stretching the timeframe to count benefits to future humans or, like Muir, they could expand the reference class of beneficiaries by including plants, animals, and nature itself. In the current context, however, maximizing arguments, while clearly retaining a background role, are pushed off center stage by questions of fairness and equitable distribution of benefits among individuals and groups. This shift in context and value considerations is all tied up with the scale at which impacts are accounted and how costs are distributed, so the problem of the proper scale at which to model and address environmental problems has become an important knowledge question.

In discussing environmental problems from a policy perspective, "scale" has two "knowledge" aspects—what is the scale of the physical system that is identified as problematic, and what is the socio-political and organizational scale (or scales) at which to address the problem? Successful environmental policy outcomes can be achieved only if there are processes capable of addressing both of these aspects. The difficulty in addressing linked conservation and development goals is in part due to a limited understanding of the complexity of choices or trade-offs that need to be made and their short- and long-term implications in monastic (e.g., neoclassical) versus pluralistic (e.g., DPM) theoretical frameworks of environmental valuation.

How the treatment of process influences valuation poses some very difficult meta-theoretical questions: Does market determination of prices through arbitrage represent a fair process? Most of the economic valuation theorists will arguably agree that market transactions represent a fair process. There is intense debate among economic valuation theorists about how to assign market level prices for goods (e.g., biodiversity) or services (e.g., carbon based ecosystem services) for which natural markets do not exist, or for which markets fail to internalize the cost or benefit externalities. A large range of Contingent Valuation (CV) methodologies have been proposed and implemented to "prize" these non-marketed goods and services and complete the valuing of "indirect" effects in a total economic valuation system. Social valuation theorists, on the other hand, argue against the assumption that markets represent a fair process. Governments are required to establish boundary rules for the robust functioning of markets processes, and, in many cases governments need to intervene in the markets to internalize the externalities, which opens up a political Pandora's Box in the determination of a fair process to undertake the valuation in a system. The social valuation theorists consider democracy and broader representation of all stakeholders in the valuation process more important than the "blind" pursuit of market prices.

The question of power in valuation theories is of prime importance due to the social and political nature of placing values on non-market goods and services, or even questioning the value bestowed by market prices. The behavioral decision theorists analyze cognitive structures of human decision-makers to understand how knowledge is used as a powerful tool to influence valuation of market and non-market goods and services. This is exactly what I mean by politics of knowledge in climate governance that assigns relatively more weight to "markets" and "human values" than "non-market" policy mechanisms and "non-human" values.

The treatment of (1) commensurability, (2) consequentialism, (3) scale, (4) process and (5) power in a valuation theory is critical for analyzing the value trade-offs in real world decision-making contexts, such as the ones requiring post-Kyoto climate governance and policy decision-making. The methodological tools engendered by the wide variety of valuation theories need to be applied in their proper theoretical context for eliciting value trade-offs. Cost-benefit analysis can, for example, enable calculation of opportunity costs and benefits for pursuing one decision alternative as opposed to some other. Whether all costs and benefits have been accounted for (the completeness issue), whether commensurability transformative measure is adequate (the commensurability issue), or whether appropriate boundaries have been chosen to implement cost-benefit analysis, are some of the major limitations that cannot be addressed without considering issues of process and power in a fair implementation of cost-benefit analysis or any other valuation tool. A more specialized application of conjoint analysis, for example, to estimate the value of carbon protection will beg the questions of who is surveyed, when, where, how often, who constructed the contingent scenarios, which scenarios were presented first, which scenarios were not included in the analysis, what were the valuation choices, what discount rates were used, and so on. Similar limitations apply when other valuation tools, such as multiple criteria decision analysis (MCDA), citizen advisory groups, Delphi panels, or targeted focus groups are used to elicit value trade-offs.

In contrast with neo-classical theory of total economic valuation, ecological imperialism theory of environmental valuation (Perelman 1999, 2003; Clark and Foster 2009), rights based valuation theories and DPM valuation theories argue for the existence and acknowledge of pluralistic value systems that are typically incommensurate. Despite the existence of such alternate theories, it has been argued that interdisciplinary fields of inquiry, such as Ecological Economics, suffer from the dominance of neo-classical economists' theory of total economic valuation. Norton and Noonan (2007: 665), for example, state:

> What worries us is that the current enthusiasm for ecosystem service methods (used in tandem with contingent valuation methods) has locked the rhetoric of environmental evaluation in a very monistic, utilitarian, and economic vernacular that leaves little or no room for other social scientific methods, or for appeal to philosophical reasons or theological ideals. It also discourages a more profound reexamination of how one might create a rational process of policy evaluation that truly takes into account both economic and ecological impacts of our decisions. At the very least, enthusiastic pursuit of monistic analysis serves to distract ecological economics from genuine development.
>
> (Norton and Noonan 2007: 665)

Norton and Noonan provided persuasive arguments to reject the "monistic, utilitarian" theory of valuation. Instead, they argued for the development of "a new, pluralistic, multi-scalar, and multi-criteria method of evaluating anthropogenic changes to natural and social systems" (Norton and Noonan 2007: 665).

Norton and Noonan distinguish between monistic and pluralistic theories of valuation by suggesting that

> monistic approaches to evaluation attempt to represent all environmental value in one framework of analysis—such as utilitarianism, cost–benefit analysis, or rights theory. Pluralistic theories, on the other hand, do not attempt to enforce a universal vocabulary upon the discourse of environmental value.
>
> (Norton and Noonan 2007: 666)

Further, they argue that

> trying to force all values at issue into a single, monistic framework leads to a politics of ideology and exclusion, as interest groups that define the problem differently struggle to gain control of public discourse and enforce the methodology that yields "one right answer." Issues of value formulation that should be discussed openly are hidden in bureaucratic decisions concerning "appropriate" discount rates, for example. Recognizing multiple values and multiple vernaculars, encouraging open discussion of values–pluralism—can lead to negotiation and reformulation of problems as people develop new, sometimes more similar, "mental models" of problem situations.
>
> (Norton and Noonan 2007: 666)

Disaffection with "monistic" and "utilitarian" theories of valuation has also been expressed in environmental valuation literature as a conflict between "commensurate" versus "non-commensurate" valuation theories. Martinez-Alier *et al.*, while responding to the issue of how value conflicts are to be resolved, state:

> one approach that has its roots in utilitarianism is that which attempts resolution through the use of a common measure through which different values can be traded off one with another: monetary measures are the most commonly used measure invoked for this purpose. The approach assumes the existence of value commensurability. Is that assumption justified? In the following we argue that it is not.
>
> (Martinez-Alier *et al.* 1998: 278)

In contrast, incommensurate values imply the non-existence of a common scale of measurement:

> Incommensurability, i.e., the absence of a common unit of measurement across plural values, entails the rejection not just of monetary reductionism but also any physical reductionism (e.g., eco-energetic valuation). However it does not imply incomparability. It allows that different options are weakly comparable, that is comparable without recourse to a single type of value.
>
> (Martinez-Alier *et al.* 1998: 280)

While arguing for value pluralism, which implies non-commensurate values, Martinez-Alier *et al.* (1998: 281) recommend the use of "multi-criteria framework as a paradigm for the whole field of ecological economics (both macro, micro and in project evaluation)." More importantly, Martinez-Alier *et al.* emphasize that they

> *do not believe in algorithmic solutions of multicriteria problems.* In our view, multicriteria methods useful for environmental policy must offer a consistent framework aimed at helping the structuring of the problem and

the evolution of the "decision process", so that "soft" approaches such as for example, discursive ethics (O'Hara, 1996) can more easily be implemented.

(Martinez-Alier *et al.* 1998: 281, italics in original)

Since this call for action by Martinez-Alier *et al.* (1998) for a deliberative/discursive multi-criteria policy evaluation framework, a number of studies have been published that demonstrate the applicability of a non-monistic, value pluralistic, multi-criteria theory of valuation with a Habermasian deliberative bent of communicative action (Martinez-Alier 2001; Wilson and Howarth 2002; Howarth and Wilson 2006; Klauer *et al.* 2006; Messner *et al.* 2006; Munda 2006; Norese 2006; Proctor and Dreschler 2006; Renn 2006; Stagl 2006; Van Den Hove 2006). This body of literature has emerged in parallel to the deliberative value focused decision analytic models (Keeney 1988; Keeney 1992; Gregory and Keeney 1994; Keeney 1996a; Keeney and McDaniels 1999). Kiker *et al.* 2005 present a broad review of studies that involve the application of multiple criteria decision-making models for environmental decision-making. Major limitations of deliberative multi-criteria evaluation methods are discussed by Hisschemöller and Hoppe (1995); Pellizzoni (2001); Shim *et al.* (2002); Stirling (2006); and Wittmer *et al.* (2006).

From a meta-theoretical standpoint, as argued above, neo-classical theory of economic valuation could not be simply privileged over other valuation theories. However, the marketization of climate governance, as evidenced in the case of Kyoto and fledgling post-Kyoto policy mechanisms such as REDD+ and CDM, appears to be based on the foundational valuation principles of neo-classical theory of total economic valuation. Such politics of knowledge disrespects other value systems and knowledge arenas. More so, I would like to claim that such politics of knowledge has led to the failure of Kyoto Protocol. In a post-Kyoto climate governance regime, alternative, value pluralistic theories of valuation as well as knowledge must be acknowledged and respected. I explore this politics of knowledge in the context of two pressing post-Kyoto policy designs problems: GHG emission entitlements across the states (in the next section) and tropical forest governance and REDD+ design (in the final section).

GHG emission entitlements

The common but differentiated "equity" principle adopted in UNFCCC negotiations for allocation of carbon emission entitlements at the international scale is expected to result in some form of grandfathering, as it happened in the case of the Kyoto Treaty. While there are a plethora of proposals for designing policy architecture to allocate GHG emission allowances, as discussed in Chapter 1, here I aim to demonstrate how the choice of a particular knowledge framework can privilege present generations over the future generations, or rich industrialized countries over the poor industrializing countries. In particular, I am proposing that the GHG emission entitlements proposed in different policy architectures can be compared with the "baseline" GHG per capita decision heuristic. Assuming that GHG per capita emission allocations represent a specific "southern" knowledge perspective on GHG emission entitlements across the human generations, the actually negotiated targets can be compared and evaluated vis-à-vis the GHG per capita allocation targets to ascertain how different knowledge frameworks result in competing GHG emission entitlements.

The baseline "GHG per capita" knowledge framework is not acceptable to many developed countries, especially JUSCANZ countries, as their representatives in UNFCC negotiations have argued that GHG per capita emission allocations are too burdensome from an economic standpoint and that these allocations condone larger population sizes of developing countries. These reservations are probably legitimate, but at the same time, it can also be argued that environmental polluters in any environmental policy design context generally raise these types of reservations when a policy design is negotiated to regulate the pollution from the polluters. Further, these excuses by developed countries have also been used for inaction on climate change. Instead of getting bogged down in these interminable arguments and counter arguments about the role of population sizes in the allocation of GHG emission allowances, it might be better for the emergence of global cooperation on climate change to let the principle of "common but differentiated" emission reductions guide the emission allocation negotiations. However, the global climate policy community needs to move forward in terms of operationalizing the notion of "common but differentiated" emission reductions in terms of "equitable" allocation of GHG emission entitlements for designing a post-Kyoto climate treaty that avoids the mistakes of the Kyoto Treaty. Kaya (1990) identity, as shown in equation 6.1, provides a useful tool to model national level CO_2 emissions.

$$CO_2 = (CO_2/E)(E/GDP)(GDP/P)P \tag{6.1}$$

In equation 1, CO_2/E represents the carbon dioxide intensity of a country, E/GDP represents the energy intensity of a country, GDP/P represents income per capita and P represents the population of a country. This version of Kaya identity has been used by a large number of studies (Albrecht *et al.* 2002; Baksi and Green 2007; Duro and Padilla 2006; Ekins 2004; Kawase *et al.* 2006; Kwon 2005; Lise 2005; Lozano and Gutierrez 2008; O'Neill *et al.* 2002; Price *et al.* 1998; Ramanathan 2006; Wang *et al.* 2005; Yang and Schneider 1998; Zhang *et al.* 2009) to model the dynamics of country level CO_2 production, especially for comparative policy analytical purposes. Equation 6.1 can also be equivalently written in terms of CO_2 per capita, as in Equation 6.2:

$$CO_2/P = (CO_2/E)(E/GDP)(GDP/P) \tag{6.2}$$

Kaya identity can be decomposed through Shapely (1953) decomposition process to estimate the marginal effects of Kaya identity components on the changes in CO_2 per capita over multiple years in a given country. Equation 6.3 presents a Shapely decomposition of Equation 6.2.

$$d(\ln CO_2/P)/dt = d(\ln CO_2/E)/dt + d(\ln E/GDP)/dt + d(\ln GDP/P)/dt \tag{6.3}$$

This kind of Shapely decomposition provides a powerful comparative policy analysis tool to evaluate climate policy intervention effects by comparing changes in Kaya identity components before and after the climate policy interventions are introduced in various countries over time.

While Shapely decomposition of Kaya identity can provide useful information to design "with-in" country climate and energy policies, an "across-the-country" approach will require some operational methodology to empirically ground the "common but differentiated" GHG emission entitlements. Next, I present a simple statistical model to calculate CO_2 emission allowances for "across-the-country" approach under the grandfathering and per capita decision rules. Databases released by IEA, World Bank and OECD are used to estimate the variables for the statistical model. A fixed effects statistical model is estimated and parameterized to estimate CO_2 emissions (in metric tons per capita per year) for each country between 1970 and 2004.

Notably, the proposed statistical approach provides a statistical measure of the bargaining space that surrounds various ethical dimensions concerning the choice of decision rules for a fair allocation of CO_2 emissions in the post-Kyoto (post-2012) implementation period. The impact of such allocation choices is expected to be huge on global economy and environment and stakes surrounding the choice of "common but differentiated" responsibilities are fairly high because the bargaining space is not insignificant. This approach demonstrates the stakes involved in resolving a policy implementation dilemma that requires a negotiated solution.

The fundamental rationale to present this statistical model here is to articulate an alternative way of estimating carbon emission entitlements that enables comparison among different decision heuristics, which might be based upon different knowledge systems and world-views. As argued in Chapter 1, grandfathering based emission reduction targets have hegemonized environmental policy discourse. The statistical model is used here to demonstrate the differences in terms of emission entitlements such as tons of GHG/year, by comparing the difference between grandfathering and a "baseline" decision heuristic. The politics of knowledge in climate governance hinges on choosing the "baseline" decision heuristic.

In addition to the emissions per capita decision heuristic that is simulated here as a baseline for demonstration purposes, grandfathered emission reduction targets could also be compared with other "baseline" policy architectures, e.g., contraction and convergence, KISS, or high emitter profiles (e.g., Chakravartya *et al.* 2009). As an example, this statistical model estimates the U.S. target to be 163 percent (~16.07 tons of CO_2 per capita) reduction below 1990 levels under a weighted per capita decision heuristic, as explained below, while the United States had signalled in the Copenhagen Conference of Parties to set up a grandfathered target of 17 percent GHG emission reductions by 2020 below their 2005 levels (which is effectively less than 5 percent emission reductions below 1990 levels) for a post-Kyoto Treaty. This comparison implies that the potential U.S. commitment of 5 percent reduction would have provided the United States a

Table 6.1 OLS and log-linear model estimates for CO_2 per capita emissions above or below global average 1990 levels

Variables	Negotiating Block	OLS 1: CO_2/Cap	Loglinear 2: $lnCO_2/Cap$
1972		−0.86*** [0.23]	−0.28*** [0.04]
1973		−0.58*** [0.23]	−0.22*** [0.04]
1974		−0.70*** [0.23]	−0.22*** [0.04]
1975		−0.69*** [0.23]	−0.20*** [0.04]
1976		−0.56** [0.23]	−0.16*** [0.04]
1977		−0.43* [0.23]	−0.12*** [0.04]
1978		−0.34 [0.23]	−0.09** [0.04]
1979		−0.09 [0.23]	−0.05 [0.04]
1980		−0.09 [0.23]	−0.04 [0.04]
1981		−0.17 [0.23]	−0.06 [0.04]
1982		−0.20 [0.23]	−0.06 [0.04]
1983		−0.25 [0.23]	−0.06 [0.04]
1984		−0.17 [0.23]	−0.06 [0.04]
1985		−0.12 [0.23]	−0.05 [0.04]
1986		−0.03 [0.23]	−0.04 [0.04]
1987		0.02 [0.23]	−0.01 [0.04]
1988		0.07 [0.23]	0.00 [0.04]
1989		0.13 [0.23]	0.01 [0.04]
1991		0.04 [0.23]	0.00 [0.04]
1992		0.08 [0.23]	0.01 [0.04]
1993		0.12 [0.23]	0.01 [0.04]
1994		0.17 [0.23]	0.02 [0.04]
1995		0.17 [0.23]	0.06 [0.04]
1996		0.33 [0.23]	0.10** [0.04]
1997		0.33 [0.23]	0.11 [0.04]
1998		0.35 [0.23]	0.12 [0.04]
1999		0.35 [0.23]	0.13 [0.04]
2000		0.40* [0.23]	0.14 [0.04]
2001		0.43* [0.23]	0.16 [0.04]
2002		0.47** [0.23]	0.17 [0.04]
2003		0.65*** [0.23]	0.20 [0.04]
2004		0.75*** [0.23]	0.22 [0.04]
Albania		−2.23*** [0.42]	−0.96*** [0.07]
Algeria	G77 OPEC	−2.10*** [0.42]	−0.83*** [0.07]
Angola	G77 OPEC	−3.59*** [0.42]	−2.45*** [0.07]
Argentina	G77	−0.59 [0.42]	−0.16** [0.07]
Australia	JUSCANZ	10.72*** [0.42]	1.31*** [0.07]
Austria	EU	3.53*** [0.42]	0.64*** [0.07]
Bahrain	G77/OPEC	17.57*** [0.42]	1.69*** [0.07]

continued

Table 6.1 Continued

Variables	Negotiating Block	OLS 1: CO_2/Cap	Loglinear 2: $lnCO_2/Cap$
Bangladesh	G77	−3.81*** [0.42]	−3.56*** [0.07]
Belgium	EU	7.73*** [0.42]	1.08*** [0.07]
Benin	G77	−3.82*** [0.42]	−3.65*** [0.07]
Bolivia	G77	−3.10*** [0.42]	−1.57*** [0.07]
Botswana	G77	−2.26*** [0.46]	−0.83*** [0.08]
Brazil	G77	−2.53*** [0.42]	−1.04*** [0.07]
Brunei Darussalam	G77/OPEC	8.67*** [0.42]	1.10*** [0.07]
Bulgaria	G77	3.65*** [0.42]	0.64*** [0.07]
Cameroon	G77	−3.75*** [0.42]	−3.08*** [0.07]
Canada	JUSCANZ	12.34*** [0.42]	1.42*** [0.07]
Chile	G77	−1.59*** [0.42]	−0.56*** [0.07]
Taipei	G77	1.73*** [0.42]	0.25*** [0.07]
China	G77	−2.05*** [0.42]	−0.79*** [0.07]
Colombia	G77	−2.65*** [0.42]	−1.13*** [0.07]
Congo	G77	−3.62*** [0.42]	−2.60*** [0.07]
Costa Rica	G77	−2.96*** [0.42]	−1.42*** [0.07]
Democratic Republic of Congo	G77	−3.86*** [0.42]	−3.97*** [0.07]
Cuba	G77 AOSIS	−1.39*** [0.42]	−0.45*** [0.07]
Cote d'Ivoire	G77	−3.60*** [0.42]	−2.50*** [0.07]
Cyprus	EU	1.60*** [0.42]	0.29*** [0.07]
Czech Republic	EU	10.28*** [0.42]	1.27*** [0.07]
Democratic People's Republic of Korea	G77	0.88** [0.42]	0.16** [0.07]
Denmark	EU	6.99*** [0.42]	1.02*** [0.07]
Dominican Republic	G77 AOSIS	−2.61*** [0.42]	−1.14*** [0.07]
Ecuador	G77 OPEC	−2.66*** [0.42]	−1.16*** [0.07]
Egypt	G77	−2.71*** [0.42]	−1.22*** [0.07]
El Salvador	G77	−3.38*** [0.42]	−2.02*** [0.07]
Eritrea	G77	−4.13*** [0.57]	−3.14*** [0.10]
Ethiopia	G77	−3.89*** [0.42]	−4.55*** [0.07]
Finland	EU	6.91*** [0.42]	1.01*** [0.07]
Former USSR	G77	5.92*** [0.42]	0.90*** [0.07]
Former Yugoslavia	G77	0.53 [0.42]	0.11 [0.07]
France	EU	3.07*** [0.42]	0.57*** [0.07]
Gabon	G77	−2.46*** [0.42]	−1.03*** [0.07]
Germany	EU	8.09*** [0.42]	1.11*** [0.07]
Ghana	G77	−3.72*** [0.42]	−2.94*** [0.07]
Gibraltar	G77	3.67*** [0.42]	0.47*** [0.07]
Greece	EU	1.98*** [0.42]	0.37*** [0.07]
Guatemala	G77	−3.41*** [0.42]	−2.04*** [0.07]
Haiti	G77 AOSIS	−3.81*** [0.42]	−3.48*** [0.07]
Honduras	G77	−3.41*** [0.42]	−2.03*** [0.07]
Hong Kong, China	G77	0.60 [0.42]	0.08 [0.07]
Hungary	EU	2.76*** [0.42]	0.52*** [0.07]
Iceland	JUSCANZ	3.52*** [0.42]	0.64*** [0.07]
India	G77	−3.29*** [0.42]	−1.87*** [0.07]
Indonesia	G77	−3.19*** [0.42]	−1.83*** [0.07]

continued

Variables	Negotiating Block	OLS 1: CO_2/Cap	Loglinear 2: $lnCO_2/Cap$
Iraq	G77 OPEC	−1.27*** [0.42]	−0.44*** [0.07]
Islamic Republic of Iran	G77 OPEC	−0.62 [0.42]	−0.23*** [0.07]
Ireland	EU	4.53*** [0.42]	0.75*** [0.07]
Israel	JUSCANZ	2.77*** [0.42]	0.51*** [0.07]
Italy	EU	2.72*** [0.42]	0.52*** [0.07]
Jamaica	G77 AOSIS	−0.79* [0.42]	−0.24*** [0.07]
Japan	JUSCANZ	4.26*** [0.42]	0.73*** [0.07]
Jordan	G77	−1.58*** [0.42]	−0.58*** [0.07]
Kenya	G77	−3.65*** [0.42]	−2.61*** [0.07]
Korea	G77	1.29*** [0.42]	0.13* [0.07]
Kuwait	G77 OPEC	18.75*** [0.42]	1.74*** [0.07]
Lebanon	G77	−0.91** [0.42]	−0.32*** [0.07]
Libya	G77 OPEC	1.90*** [0.42]	0.34*** [0.07]
Luxembourg	EU	24.30*** [0.42]	1.93*** [0.07]
Malaysia	G77	−1.20*** [0.42]	−0.49*** [0.07]
Malta	EU	0.67 [0.42]	0.07 [0.07]
Mexico	G77	−0.73* [0.42]	−0.22*** [0.07]
Morocco	G77	−3.12*** [0.42]	−1.60*** [0.07]
Mozambique	G77	−3.81*** [0.42]	−3.52*** [0.07]
Myanmar	G77	−3.80*** [0.42]	−3.34*** [0.07]
Namibia	G77	−3.22*** [0.55]	−1.41*** [0.09]
Nepal	G77	−3.88*** [0.42]	−4.43*** [0.07]
Netherlands	EU	6.88*** [0.42]	1.01*** [0.07]
New Zealand	JUSCANZ	2.45*** [0.42]	0.47*** [0.07]
Nicaragua	G77	−3.32*** [0.42]	−1.86*** [0.07]
Nigeria	G77 OPEC	−3.60*** [0.42]	−2.52*** [0.07]
Norway	JUSCANZ	3.07*** [0.42]	0.57*** [0.07]
Oman	G77	0.46 [0.42]	−0.26*** [0.07]
Other Africa		−3.81*** [0.42]	−3.41*** [0.07]
Other Latin America		0.19 [0.42]	0.04 [0.07]
Other Asia		−3.52*** [0.42]	−2.34*** [0.07]
Pakistan	G77	−3.45*** [0.42]	−2.16*** [0.07]
Panama	G77	−2.41*** [0.42]	−0.96*** [0.07]
Paraguay	G77	−3.46*** [0.42]	−2.16*** [0.07]
Peru	G77	−2.90*** [0.42]	−1.34*** [0.07]
Philippines	G77	−3.23*** [0.42]	−1.73*** [0.07]
Poland	EU	5.82*** [0.42]	0.90*** [0.07]
Portugal	EU	−0.36 [0.42]	−0.18*** [0.07]
Qatar	G77 OPEC	30.24*** [0.42]	2.13*** [0.07]
Romania	G77	2.34*** [0.42]	0.44*** [0.07]
Saudi Arabia	G77 OPEC	5.75*** [0.42]	0.78*** [0.07]
Senegal	G77	−3.64*** [0.42]	−2.57*** [0.07]
Singapore	G77 AOSIS	3.32*** [0.42]	0.53*** [0.07]
Slovak Republic	EU	4.97*** [0.42]	0.81*** [0.07]
South Africa	G77	3.43*** [0.42]	0.62*** [0.07]
Spain	EU	1.44*** [0.42]	0.29*** [0.07]
Sri Lanka	G77	−3.62*** [0.42]	−2.57*** [0.07]
Sudan	G77	−3.75*** [0.42]	−3.07*** [0.07]

continued

Table 6.1 Continued

Variables	Negotiating Block	OLS 1: CO_2/Cap	Loglinear 2: $lnCO_2/Cap$
Sweden	EU	3.52*** [0.42]	0.62*** [0.07]
Switzerland	JUSCANZ	2.12*** [0.42]	0.43*** [0.07]
Syria	G77	−1.79*** [0.42]	−0.65*** [0.07]
Thailand	G77	−2.48*** [0.42]	−1.20*** [0.07]
Togo	G77	−3.79*** [0.42]	−3.34*** [0.07]
Trinidad and Tobago	G77 AOSIS	5.65*** [0.42]	0.84*** [0.07]
Tunisia	G77	−2.56*** [0.42]	−1.09*** [0.07]
Turkey	JUSCANZ	−1.83*** [0.42]	−0.66*** [0.07]
United Arab Emirates	G77 OPEC	18.07*** [0.42]	1.65*** [0.07]
United Kingdom	EU	5.83*** [0.42]	0.91*** [0.07]
United Republic of Tanzania	G77	−3.86*** [0.42]	−3.91*** [0.07]
United States of America	JUSCANZ	16.07*** [0.42]	1.63*** [0.07]
Uruguay	G77	−2.37*** [0.42]	−0.94*** [0.07]
Venezuela	G77 OPEC	1.38*** [0.42]	0.30*** [0.07]
Vietnam	G77	−3.56*** [0.42]	−2.40*** [0.07]
Yemen	G77	−3.42*** [0.42]	−2.11*** [0.07]
Zambia	G77	−3.52*** [0.42]	−2.39*** [0.07]
Zimbabwe	G77	−2.76*** [0.42]	−1.23*** [0.07]

Notes
* shows significance at 0.01 level and ** shows significance at 0.05 level and *** shows significance at 0.001 level. Numbers in brackets ( ) are standard errors.

grandfathered targets that are being negotiated in UNFCCC process for post-Kyoto Treaty. For example, in OLS1 model, the United States CO_2/capita emissions are predicted to be 16.07 tons higher than the 1990 global average, which is very similar to the Loglinear1 model prediction, i.e., the U.S. emissions are 163 percent higher than the 1990 global average. If CO_2/capita emission rule (with the baseline of 1990 emissions) were to be imposed on the United States, this finding implies that the United States must agree to a binding reduction of ~163 percent lower CO_2/capita emissions than 1990 levels. In contrast, this can be compared with the actual negotiation data, whereby the United States has at best offered a 17 percent reduction below 2005 baseline levels! Further, 17 percent reduction on a 2005 baseline is well below the predicted 163 percent reduction on a 1990 baseline if the post-Kyoto Treaty enforces CO_2/capita emission decision heuristic. This raises the politics of knowledge in climate governance: as the choice of one model or decision heuristic over another could change the climate policy outcomes, especially emission entitlements could vary drastically by such choices. The qualitative discourse analysis of UNFCCC negotiations confirms that grandfathering based decision heuristic has so far dominated the GHG allocation issues. Similar, country-parameter results on other countries presented in Table 6.1 could be compared with their UNFCCC offers of emission reduction targets, as briefly explained below at the level of negotiating blocs and countries within those blocs.

In general, both the models show that OPEC countries, which are opposed to any binding GHG emission reductions, typically emit very high CO_2 per capita, such as Bahrain averages 17.57 tons (169 percent), Brunei Darussalam averages 8.67 tons (110 percent), Kuwait averages 18.75 tons (174 percent), Saudi Arabia averages 5.75 tons (78 percent), and UAE averages 18.07 tons (165 percent) above the 1990 CO_2/capita world average. If a binding per capita decision heuristic is applied, Bahrain, Brunei, Kuwait, Saudi Arabia and UAE will be required to respectively purchase 17.57, 8.67, 18.75, 5.75 and 18.07 tons of CO_2 per capita entitlements on an annual basis according to this model for the period 1971–2004. In contrast with these findings, the OPEC countries have argued that they must be compensated for any losses in their oil-driven revenues, were UNFCCC to impose any binding GHG reductions!

The JUSCANZ countries also typically emit CO_2 per capita well above global average. The CO_2 per capita for Japan averages about 4.26 tons (73 percent), the U.S. averages 16.07 tons (163 percent), Canada averages 12.34 tons (142 percent), Australia averages 10.72 tons (131 percent) and New Zealand averages 2.45 tons (47 percent) above the 1990 world average CO_2/capita emissions. If in a post-Kyoto Treaty, grandfathering is applied again as it was done for the Kyoto Treaty, where Japan and Canada agreed to 6 percent reduction below their 1990 levels, the negotiated grandfathered percent can be compared with the per capita entitlement calculated from this model. So, this model estimates that Japanese and Canadians should be required to reduce their per capita CO_2 emissions by 73 percent and 142 percent respectively; however, it is expected that grandfathered targets negotiated by these two countries could be significantly lower than predicted by per capita model as demonstrated in the 6 to 8 percent reductions committed under Kyoto Protocol. Similarly, EU-25 countries also typically produce higher than world average CO_2 per capita emissions with few exceptions (e.g., Portugal).

The G-77/developing countries typically produced below world average CO_2 per capita emissions. Brazil, China and India, for example, respectively produced 2.53 tons (104 percent), 2.05 tons (79 percent) and 3.29 tons (187 percent) lower CO_2 per capita emissions than the 1990 world average CO_2/capita emissions. Under CO_2/capita decision heuristic, the developing countries can be given the corresponding allowance to increase their emissions, but once they reach the targeted global average, they can be graduated and required to purchase additional CO_2 per capita emission entitlements. This position is however not acceptable to OPEC and JUSCANZ countries.

The common but differentiated principle adopted in UNFCCC negotiations for allocation of carbon/GHG emission entitlements at the international scale is, in a post-Kyoto regime, expected to result in some form of grandfathering, as it happened in the case of the Kyoto Treaty, due to naturalization of grandfathering in the UNFCCC driven global climate change discourse. While there are a plethora of proposals for designing policy architectures to allocate CO_2 and other non-carbon GHG emission allowances, an important question is how to evaluate these policy architectures on various criteria of evaluation. Earlier in Chapter 1,

I evaluated policy architectures on the criteria of binding versus non-binding allocation of GHG emissions. I have argued that carbon emission entitlements negotiated through UNFCCC process could be compared with different policy architectures to estimate the benefits that high emitting countries are potentially drawing by naturalizing grandfathering as an "inevitable" choice for a decision heuristic. Assuming that per capita emission allocations represent "baseline" decision heuristic, the UNFCCC negotiated targets can be compared and evaluated vis-à-vis the "baseline" decision heuristic.

The CO_2/capita baseline decision heuristic is probably not acceptable to many developed countries, especially JUSCANZ countries, as their representatives in UNFCCC negotiations have argued that per capita emission allocations are either too burdensome from economic standpoint or that these allocations condone larger population sizes of developing countries. These reservations are probably legitimate and raise important politics of knowledge questions in terms of pinning down historical responsibility versus the future right of development for developing countries. Despite these statistical and ethical dilemmas, the global climate policy community needs to move forward in terms of operationalizing the notion of "common but differentiated" emission reductions by allocating GHG emission entitlements for designing a post-Kyoto climate treaty.

On marketization of carbon governance, it is pertinent to state that markets can be created by the invention of a new product or service, a patent, or when someone or a small group of people can have exclusive ownership of a certain good or service. What carbon markets do is give ownership of a small portion of the waste absorption capacity to a specific industry. Mechanisms like CDM permit "industrialized countries to earn emission credits through investment in sustainable development projects that reduce overall emissions in developing countries" (Sharma 2011). The language here is clear but the underlying message is slightly more complicated. This means that large industrialized nations can invest money into less industrialized nations to reduce their emissions and therefore earn credits that allow the industrialized nation to pollute more. There is a disconnect between the size of the country, the amount they pollute, and the political powers that govern them both. Though in theory carbon markets can slow down unsustainable development by putting a price on pollution, it does not reverse or call for a significant change in consumption behavior. The fact is that "greenhouse gases are a global pollutant without local effects, the location of the emissions does not matter" (Michaelowa and Schmidt 1997: 45). This means that policy-makers have to look at emissions as total gross emissions around the world; that reduction in emissions in one part of the world does not give another part of the world carte blanche to pollute as they please.

There are many factors, mainly surrounding economic reasons, why inequity plays such a huge role in determining the efficiency of carbon markets. As with any other markets, prices often fluctuate based on speculation, and in the case of carbon, limited knowledge and uncertainty often causes prices to fluctuate (Michaelowa and Schmidt 1997). Reductions in prices can "allow other nations to inexpensively buy credits rather than focus on mitigation efforts" (Evans *et al.*

2000: 322). Putting a direct price on emissions creates the possibility for them to either be cheap or expensive. Market forces are efficient for some things like deciding the price of jeans and t-shirts, but with carbon there is a little more at stake, and we cannot simply rely on global markets to regulate pollution properly. Carbon emissions and waste absorption capacity are naturally non-excludable goods (Daly and Farley 2010). Turning them into excludable goods by putting a price on them allows those goods to take a life of their own. Markets function by allocating resources to those who value them the most, and value is determined by the price someone is willing to pay for it rather than what it is actually worth. Those who have the most value for carbon emission credits are clearly going to be rich and powerful corporations, so why should we structure a mechanism that directly allocates more power to those that already have too much? The use of carbon markets to solve the problems of environmental degradation means putting all faith into the theoretical functions of markets that, as stated before, seem to favor those that have the most amount of financial capital, and that, in the first place, contributed most to cause the global climate change problem.

Tropical forest governance and foreign direct investments

Broadly, the emphasis on marketization of climate governance in the context of the complexity in coupled natural and human systems raises issues of politics of knowledge, politics of scale and politics of ideology across disciplines that require the immediate attention of international policy-makers and negotiators. In the specific context of international climate and economic development policies, the politics of knowledge lens analyzes whose knowledge "hegemonizes" the design of climate change mitigation and adaptation policies in developing countries (MacMynowski 2007). The same may be said for the role of multilateral institutions such as the World Bank and the International Finance Corporation in establishing the rules of the game in FDI (Daniel and Mittal 2010). For example, with FDI in land-based enterprises, one stream of knowledge emphasizes the local and national economic benefits, while the other stream of knowledge emphasizes the social and environmental costs associated with the displacement of forests and customary land users. The politics of knowledge lens sheds light on such policy disputes and elucidates the politicized nature of climate and development science. In particular, the role of policy narratives and knowledge hegemony during international negotiations could be assessed through politics of knowledge analytical lenses. Such counter-narratives also point to the crucial importance of grounding decisions in a clear understanding of social, economic and environmental impacts and trade-offs.

Current trends in tropical deforestation pose daunting challenges for mitigating global climate change and conserving biodiversity, with mixed effects on poverty and food security. Using forest inventory data and long-term ecosystem carbon studies, Pan *et al.* (2010) estimated a total forest sink of 2.4 ± 0.4 petagrams of carbon per year ($PgC year^{-1}$) globally for 1990 to 2007. Pan *et al.* (2010) also estimated a source of $1.3 \pm 0.7 PgC year^{-1}$ from tropical land-use change, consisting of a gross tropical deforestation emission of $2.9 \pm 0.5 \, \mathrm{PgCyear^{-1}}$ partially compensated by a carbon sink in tropical forest regrowth of $1.6 \pm 0.5 \, \mathrm{PgCyear^{-1}}$. Together, the carbon fluxes comprise a net global forest sink of $1.1 \pm 0.8 \, \mathrm{PgCyear^{-1}}$, with tropical estimates having the largest uncertainties. The drivers of tropical deforestation operate at multiple spatio-temporal scales, ranging from international commodity trade and investments at a global scale to national agricultural production and local economic development needs.

In a recent study, Defries *et al.* (2010) used satellite-based estimates of forest loss for 2000 to 2005 to assess economic, agricultural and demographic correlates of tropical deforestation across 41 countries in the humid tropics. They found that forest loss is positively correlated with urban population growth and exports of agricultural products for this time period. In contrast, they also found that rural population growth is not associated with forest loss, indicating the importance of urban-based and international demands for agricultural products as drivers of deforestation. Thus, the strong trend in movement of people to cities in the tropics is, counter-intuitively, likely to be associated with greater pressures for clearing tropical forests. Defries *et al.* (2010) therefore recommended that policies to reduce deforestation among local, rural populations, such as REDD+, will not address the main cause of deforestation in the future. Rather, efforts need to focus on reducing deforestation for industrial-scale, export-oriented agricultural production, concomitant with efforts to increase yields in non-forested lands to satisfy demands for agricultural products. Defries *et al.* (2010) and Hirsch *et al.* (2011) represent a growing body of literature that is calling for a broader analysis to address drivers of tropical deforestation in designing international climate policy mechanisms such as REDD+, than merely focusing on providing payment for carbon ecosystem services to national governments or local populations.

The global research community needs to analyze the emergence, design and administration of REDD+ in the face of dynamic foreign direct investment (FDI) in-flows into tropical forest countries. It could be hypothesized: *While REDD+ is designed to promote forest conservation, FDI is a dynamic driver of forest conversion and greenhouse gas emissions from land-use change and forestry (LUCF).* Clearly, REDD+ and FDI have opposite effects on the stocks of tropical forests in tropical countries, and carbon and biodiversity contained therein (Butler *et al.* 2009; Cotula *et al.* 2009; Kissinger 2011; Wells and Paoli 2011). At the same time, whether or not a location is the subject of REDD+ projects or FDI initiatives will also have significant impacts on the types of livelihood practices and local conditions people experience. From the perspective of the dynamics of coupled natural and human systems, a compelling question is which of these effects (balancing feedback effect of REDD+ to reduce deforestation or reinforcing feedback effect of FDI to enhance deforestation) will dominate tropical forested landscapes?

The emergence of REDD+ as a multi level climate policy mechanism since UNFCCC COP 15 Copenhagen negotiations has gathered renewed momentum in the wake of failure in other climate mitigation policy mechanisms, such as binding national commitments (e.g., Okereke *et al.* 2009, Zia and Koliba 2011). Some studies have forecasted that REDD+ could end tropical deforestation in individual countries such as Brazil (e.g., Nepstad *et al.* 2009), while others have developed global scale simulation models to predict that properly designed REDD+ could effectively eliminate "international leakage" and reduce global tropical deforestation by more than 90 percent at an annual cost of $30 billion per year (e.g., Strassburg *et al.* 2009). Despite these very positive potential benefits of REDD+ for carbon sequestration, the emergence of REDD+ must be analyzed in the broader context of marketization of carbon as the analytical framework for carbon governance (Newell and Patterson 2010), the dynamic context of market-based drivers of deforestation such as FDI (Wells and Paoli 2011), and the complexity associated with the changes in coupled natural and human systems (Sitch *et al.* 2003; Liu *et al.* 2007; Willis and Bhagwat 2009). The creation of international market-based climate policies, such as REDD+, are indicative of new forms of *carbon governance* that emphasize quantitative assessment and monitoring of environmental goals through the use of greenhouse gas inventories and other carbon-centered policy mechanisms (Rice 2010; Wilkie *et al.* 2010). Many social scientists have critically examined the rise of carbon economies, suggesting that their success may be limited in several ways (Lovell *et al.* 2009). Bailey and Wilson (2009) argue, for example, that the commodification of carbon through mechanisms such as emissions trading and voluntary carbon offsets works to incorporate climate policy into the "prevailing paradigm" of neoliberalism, which frames climate protection in terms of economic goals at the expense of other priorities. More importantly, this reinforcement of carbon capitalism and neoliberalism in climate governance may, indeed, prevent more environmentally and socially just forms of economic development from taking hold, while profit-driven approaches to climate governance may actually work to protect the capitalist interest that set the stage for climate change to occur in the first place (Bumpus and Liverman 2008, Lovell *et al.* 2009).

At the same time, FDI in forest-rich tropical countries and landscapes, which represents global capital flows to expand commodity production and extraction, is facilitated by bilateral and multilateral institutions as well as host country governments seeking to increase investment into rural areas in the hopes of increasing employment, improving macro-economic indices (GDP, balance of trade) and generating government revenue. The United Nations Conference on Trade and Development (UNCTAD) work program on international investment agreements, most notably technical assistance provided to developing countries and the facilitation of intergovernmental discussions, plays a fundamental role in shaping the international governance architecture of FDI. On the other hand, REDD+ has emerged as a flexible market mechanism under the aegis of UNFCCC negotiations. From a broader institutional design perspective, markets represent one specific way of designing global to local policies. The marketization of carbon governance, as manifested in REDD+ design, poses significant challenges if other market mechanisms, such as FDI, are ignored in the design of multi level climate policy mechanisms. Figure 6.1 shows the estimated relationship of LUCF induced GHGs against FDI for six tropical countries—Brazil, Indonesia, Ghana, Peru, Vietnam and Tanzania—for the GHG inventory data that was submitted to UNFCCC since 1990. Clearly, higher FDI implies higher LUCF driven GHGs.

Due to FDI and international trade in agriculture, as evidenced through increasing land grabs in Africa, Latin America and Asia, the developed countries have set up a global trading regime in the form of WTO that will continue to challenge the fundamental premises of a successful REDD+ policy. One challenge concerns the increased land use pressure on carbon-poor ecosystems with high biodiversity outside REDD+ areas: the choice of forest areas to be protected based on the potential allocation of emission reduction benefits might not coincide with areas of high biodiversity value. On the contrary, since biodiversity hotspots can have high land use conversion rates, conserving these areas might be comparatively expensive for REDD+. As a consequence, REDD+ protection

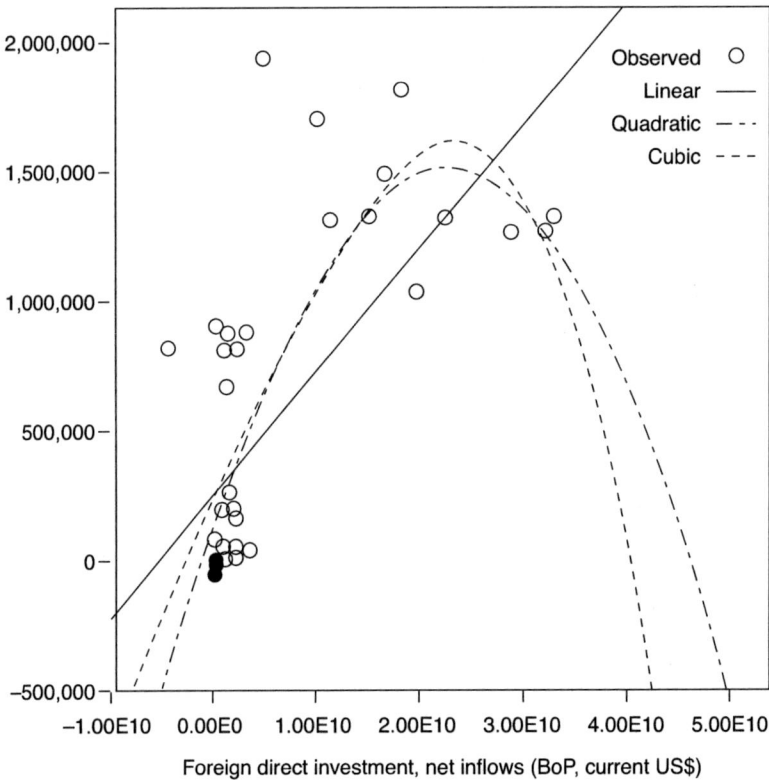

Figure 6.1 The relationship between FDI and LUCF induced GHGs in six tropical countries (data sources: UNFCCC and World Bank).

efforts might shift land use pressure to forest or non-forest areas of high biodiversity but low greenhouse gas mitigation potential, where the benefits from forest protection are comparably low (Phelps *et al.* 2011). A second challenge concerns the conversion of biodiversity-rich forest and non-forest land into plantations due to insufficient forest definitions. Under the current forest definition of the UNFCCC biodiversity-rich, natural forest could be replaced by monoculture, genetically modified or non-native tree species without being considered deforestation. This might be attractive when the greenhouse gas storage from biodiversity-poor, fast-growing plantations compensates for the greenhouse gas loss (and economic revenue from logging) of natural forest conversion.

The terms "sustainable management of forests" and "enhancement of forest carbon stocks," which were confirmed in the scope of REDD+, still need to be defined. Conversion of open savanna forests into densely covered unnatural carbon plantations might count as enhancement of forest carbon stocks or even sustainable management of forests. All these conversions into plantations could result in reduced environmental integrity compared to the previous land cover (Phelps *et al.* 2011).

A third challenge with REDD+ is the continuation of illegal logging practices in REDD+ areas due to insufficient law enforcement: Governance ineffectiveness on national and local level could contribute to illegal logging. REDD+ might reduce the incentive for local stakeholders to engage in illegal activities, when they profit from carbon payments. However, when weak law enforcement prevails, the incentive for continuing these illegal activities perpetuates (Larson 2011). Illegal logging could manifest in loss of permanence in REDD+ areas or in leakage to poorly monitored areas outside REDD+ projects.

A fourth challenge of implementing REDD+ would be land conflicts and poverty retention due to insufficient involvement of forest-dependent peoples: many studies support the expectation that REDD+ will lead to poverty reduction, since high percentages of poverty can generally be found at tropical forest margins (Larson 2011). However, if structural and governance circumstances such as land tenure security and enforcement as well as corruption reduction cannot be ensured REDD+ might not be successful in poverty reduction. The exclusion of local land users and indigenous communities in planning and revenue sharing for REDD+ measures could even lead to their further marginalization (Larson 2011).

Another challenge regarding REDD+ that is worth mentioning is the possibility of an increase in land rents and food prices due to REDD+ induced scarcity of agricultural land: REDD+ protection efforts reduce the availability of potential agricultural land, which can contribute to the increase in land rents and crop prices. The highest agricultural revenues in tropical countries can be obtained from so-called cash crops for exports (Phelps *et al.* 2011). If suitable land is occupied with high-profit cash crops, the opportunity costs for domestic agriculture, especially for low-benefit subsistence agriculture, are thus likely to exceed their economic revenue. The resulting price increases of domestic agriculture and forestry goods could disadvantage poor people, who are not able to compensate those increases or substitute their demands. Consequently, poverty and hunger could be aggravated for marginalized people in developing countries (Phelps *et al.* 2011).

Overall, governance risks such as ineffective national finance distribution, the continuation of illegal logging practices as well as the insufficient involvement of forest-dependent peoples are perceived as the major challenges for REDD+ implementation (Phelps *et al.* 2011). Many skeptics are currently arguing that REDD+, even now at its early stages, is already functioning as a form of governance, a particular framing of the problem of climate change and its solutions that validates and legitimizes specific tools and solutions while marginalizing others (Thompson *et al.* 2011). They argue that "REDD is full of assumptions about environmental governance that remain unacknowledged and therefore unexamined, creating significant challenges for productive project and policy design" (Thompson *et al.* 2011).

One of the new frontiers concerns the design of multi level institutional mechanisms based upon our understanding of cross-scale dynamics in coupled social ecological systems. In a recent speech at the annual meeting of Earth System Governance in Fort Collins, CO, Oran Young, one of the most influential scholars of our age on global policy regime theory (Young 1999, 2002), proposed that the success of global policy regimes is tied more closely to the match between the institutions and the biophysical and socioeconomic settings in which they operate than to some generic measure of how easy or hard are the problems to solve. As an institutional analyst, it appears that Oran Young has arrived at the similar conclusion as Ostrom (2007) argued for sustainability sciences to move beyond the age of panaceas.

It is not surprising that both Oran Young and Elinor Ostrom have proposed very context specific investigations of institutional dynamics in social ecological systems. In fact, Oran Young hypothesized that institutional change follows pathways that lead to one of several emergent patterns, such as progressive development, punctuated equilibrium, arrested development, diversion or even collapse. The design of a post-Kyoto climate governance regime then must be concerned about what Oran Young calls "applied institutional analysis," whereby "institutional diagnostics" are deployed as experimental means to tailor policy regimes to the needs of specific situations at different scales of the problem. In the next chapter, I elaborate such institutional diagnostics through the lens of politics of knowledge to investigate the lack of accountability in climate governance networks and the recent emergence of "adaptation fund" since the failure of mitigation policies.

7 The politics of knowledge II
Accountability and adaptation

Towards accountable and adaptive governance

Building on the politics of knowledge analysis in the previous chapter, this chapter elucidates the role of "hegemonic" knowledge on the social construction of accountability and adaptation in the global climate governance. Many specific "governance networks" are analyzed to demonstrate the power of knowledge hegemony in influencing climate policy goals to implicitly favor the rich knowledge producers at the expense of poor (or future) knowledge recipients. In the next section, the politics of knowledge lens is used to investigate various climate policy design issues pertaining to institutionalizing accountability in multi level complex governance networks that cut across public and private sector boundaries. Further, in the final section, an emergent policy mechanism known as "adaptation fund" is described and assessed in the broader context of global climate governance and sustainable development. Here, I situate the politics of climate change knowledge in the larger politics of sustainable development knowledge discourses and demonstrate that reductionist and hegemonic narratives of sustainable development and climate change adaptation, which focus on improving the lot of developing countries but ignore the past, present and future practices of developed countries in engendering the global sustainability crisis in the first place, also appear to be dominating the discourse of climate change knowledge production.

Climate governance and accountability

Innovative forms of public–public, private–private and public–private partnerships that form the basis of inter-organizational networks operating at multiple geographical scales have recently evolved to create a fragmented system of governance of climate change (Bäckstrand 2008; Biermann *et al.* 2009). These partnerships typically emerge to address "wicked" and complex public policy problems, such as global climate change, fisheries protection (Kooiman 2008) and public infrastructure provision (Vining *et al.* 2005). A growing number of studies characterize the range of partnerships as "governance networks" (Klijn 1996; Jones *et al.* 1997; Kickert *et al.* 1997; Lowndes and Skelcher 1998;

Skelcher 2005; Sørensen and Torfing 2005; Torfing 2005; Bogason and Musso 2006; Klijn and Skelcher 2007; Coen and Thatcher 2008; Koliba *et al.* 2010). A governance network is defined here as relatively stable patterns of coordinated action and resource exchanges involving policy actors crossing different social scales, drawn from the public, private or non-profit sectors and across geographic levels, who interact through a variety of competitive, command and control, cooperative, and negotiated arrangements for purposes anchored in one or more facets of the policy stream (Koliba *et al.* 2010: 60). While interdisciplinary enthusiasm for the characterization and analysis of governance networks has grown considerably, much more theoretical and empirical work remains to be done to understand how accountability is institutionalized within and across governance networks (Provan and Milward 1995; Bardach and Lesser 1996; Milward 1996; Agranoff and McGuire 2001; Papadopoulos 2003, 2007; Benner *et al.* 2004; Slaughter 2004; Frederickson and Frederickson 2006; May 2007). Zia and Koliba (2011) assess these accountability and performance management issues in the context of post-Kyoto international climate policy design and address a specific question: How can accountability be institutionalized across governance networks that are dealing with transboundary pollution problem of mitigating greenhouse gas (GHG) emissions at multiple spatial, temporal and social scales?

Notwithstanding its significance as a milestone in global environmental policy, the Kyoto Protocol has failed on many accounts in setting up effective and accountable governance mechanisms for reducing GHGs (Cass 2006; Harrison and Sundstrom 2007). Many large emitters of GHGs did not even bother to sign the treaty, as described in terms of the so-called United States–China "suicide pact" (Romm 2007). In their paper, Zia and Koliba (2011) advance the theoretical and empirical research on climate governance through the comparative analysis of accountability in different types of climate change partnerships. Over the last two decades, a variety of network strategies have been devised to address the need to mitigate the factors that are contributing to climate change. In their analysis, Zia and Koliba (2011) argue that the complexity of these governance networks gives rise to a variety of accountability challenges. These challenges are accentuated by the range of public–public, public–private, and private–private partnership arrangements that have arisen from these efforts. They argue that each type of partnership arrangement brings with it certain kinds of accountability challenges. However, they also argue that certain accountability challenges confront all forms of international partnership arrangements being devised to mitigate climate change.

In the recent climate change governance literature efforts have been initiated to develop an accountability framework for evaluating the public–public, public–private and private–private climate change governance networks. Bäckstrand (2008) develops a process-based notion of accountability that include three accountability criteria: (1) transparency; (2) monitoring mechanisms; and (3) representation of stakeholders. While process-based criteria are important components for evaluating accountability of governance networks, it has been suggested in the broader literature on pluralistic concepts of accountability in governance networks that actors (both individuals and organizations) and outcomes should also be considered as important criteria for comparing the institutionalization of accountability in governance networks (Benner *et al.* 2004). Zia and Koliba (2011) review the relevant literature on the evolution of the concept of accountability for actors, processes and outcomes in complex governance networks and describe important features of an integrative framework that can be used to comparatively analyze accountability mechanisms across governance networks. In particular, they extend Bäckstrand's (2008) process-based model of accountability for climate change governance networks by incorporating the additional criteria of actor to actor accountability predicated on the nature of ties between them, and then relate these accountability ties to more widely adopted outcome-based forms of accountability. They propose a "Governance Network Accountability Framework" to compare democratic, market and administrative anchorages of actor accountability within and across governance networks. Zia and Koliba (2011) undertake a comparative analysis of performance outcome measures in a stratified sample of public–public, public–private and private–private climate governance networks. This comparative analysis identifies four critical international climate policy design dilemmas that confront humanity for institutionalizing accountability in a global climate governance regime. These dilemmas are related to four questions: First, how to develop consistent performance measures when different governance networks propose different measures? Second, how to incorporate scientific uncertainty into performance measures? Third, how to integrate emission entitlements across multiple space-time scales? Fourth, how to monitor and verify performance benchmarks? Finally, Zia and Koliba discuss the implications of these performance management and accountability dilemmas in the context of "international democracy-deficit," "politics of knowledge" and "inter-generational accountability" to inform the evolving negotiations on designing international climate policy in the post-Kyoto (post-2012) timeframe. Here, these discussions are revisited in the context of politics of knowledge.

The governance network accountability framework

"Accountability is traditionally defined as the obligation to give an account of one's actions to someone else, often balanced by a responsibility of that other to seek an account" (Scott 2006: 175). In essence, accountability structures arise when a certain measure of interdependency exists between those rendering account (hereafter "accountees") and those to whom accounts should be rendered (hereafter "accounters"). Zia and Koliba (2011) discuss governance as a matter of accountability, with feedback taking place as processes of rendering accounts to particular constituencies, relying on certain explicit standards and tacit norms to do so. This feedback effect of communicating performance information from accountees to accounters has also been characterized as an important feature that distinguishes performance *measurement* from performance *management* systems (Kelman 2006). "Performance management is thus seen as a potentially powerful tool to remedy underperformance in government" (Kelman 2006: 394). Applying Kelman's notion of performance management to accountability in governance networks, Zia and Koliba (2011) assert that network accountability is a system level construct involving iterative performance feedback loops between accountees and accounters (Koliba *et al.* 2010, 2011). The performance feedback loops contain the flow of information on performance measures, which include information on inputs, activities and processes, actors, outputs and outcomes across the system.

Page (2004), Posner (2002), and Behn (2001) have all noted the accountability challenges associated with governance networks, recognizing their complexity and the potential competing aims inherent to the organizations operating within them. Mashaw (2006: 118) calls for the comparison of accountability regimes operating within and across network structures in order to "evaluate their differential capacities, and perhaps articulate hybrid regimes that approximate optimal institutional designs." In cases where a governance network is comprised of non-profit and for profit organizations working with governments (e.g., most public–private partnerships), the accountability regimes historically ascribed to governments are not sufficient. According to Scott,

> conventional accountability narratives, emphasizing ex-post and hierarchical forms of accountability, with only very limited reach beyond the state actors, are unable to support the burden of providing a narrative of accountability that can legitimate governance structures involving diffuse actors and methods.
>
> (Scott 2006: 190)

Zia and Koliba (2011) present a governance network accountability framework that they have used to study accountability across complex governance networks. Their initial application of this framework focused on the accountability failures found within the response and recovery networks following landfall of Hurricane Katrina in 2005 (Koliba *et al.* 2011). Within this framework, various state and non-state actors who are perceived as accountees and accounters in governance networks engage in adaptive processes over time to select an evolving set of performance measures. In highly functioning performance management systems, performance outcomes are monitored and verified by some mechanism, some information about which is fed back to accountees and accounters for completing the loop of a performance management system (Moynihan 2008).

Zia and Koliba (2011) argue that the sustainability of accountability ties within complex governance networks is difficult to accomplish. Radin (2006: 35) warns that, "despite the attractive quality of the rhetoric of the performance movement, one should not be surprised that its clarity and siren call mask a much more complex reality." Performance management is a complicated matter within *individual* organizations, let alone inter-organizational networks. Just what amounts to effective performance within a complex governance network is a matter of perception. It has been noted how performance data and standards come about through the social construction of knowledge that is predicated on a culture of performance fostered within individual organizations (Moynihan 2008) and across complex governance networks (Frederickson and Frederickson 2006; Koliba *et al.* 2010). Gregory Bateson has noted that "the processes of perception are inaccessible; only the products are conscious" (Bateson 1988: 32). Performance data, performance measures, and ultimately, performance management is complicated by the question of whose perceptions matter? Zia and Koliba (2011) assert that, presumably, accounters are in the best or the most legitimate position to determine what it means for any social entity to "perform," and presumably, perform effectively.

Many have noted how the shift from a mono-centric system of *government* to a polycentric system of *governance* raises complex *actor* accountability challenges (Behn 2001; Posner 2002; Page 2004; Goldsmith and Eggers 2004; Pierre and Peters 2005; Scott 2006; Mashaw 2006; Mathur and Skelcher 2007). Because it can no longer be assumed that the nation-state possesses the same kind of vertical authority as traditionally ascribed to governments, governing the actors in inter-organizational networks gives rise to new accountability challenges that cannot be simply modelled through conventional mono-centric accountability systems. These challenges arise when nation-states are displaced as central actors; market forces are considered; and cooperation and collaboration is recognized as an integral administrative activity.

Table 7.1 provides an overview of the governance network accountability framework (Koliba *et al.* 2010, 2011). The framework is predicated on eight different types of accounters *to whom* accountability must be rendered in a complex governance network that includes actors from government, private and civil society organizations. These accounters, be they elected representatives, citizens, courts, shareholders, consumers, supervisors, professionals or collaborators are placed in the position of judging the performance of the agents that are being held accountable as accountees. These accountees may come from any number of different kinds of actors. Complex performance management problems arise when accounters prioritize conflicting combinations of policy goals, performance measures, and other desired procedures and outcomes in a governance network, placing value on and rendering judgment of performance differently (Gruber 1987; Radin 2006). It is also imperative that accounters are capable of or interested in fulfilling their roles, which in the case of climate governance is a serious problem as a large number of potential accounters are either future generations or non-human species that will face the consequences of climate change under business-as-usual scenarios, as predicted in the 2007 IPCC synthesis report (IPCC 2007).

Because the governance network accountability framework allows for the mingling of democratic, market and administrative factors, one can view accountability in terms of trade-offs between accountability types—be they trade-offs between democracy and market accountabilities, democracy and

Table 7.1 Accountability frames for actors in complex governance networks

ACCOUNTABILITY FRAME	Accounters (those "to whom" accounts are rendered)	Relational power	Explicit standards	Implicit norms
DEMOCRATIC	Elected representatives and Courts	Vertical over public sector	Laws, statutes, regulations	Representation of collective interests; policy goals
	Citizens and courts	Horizontal accesses to public sector organizations/elected officials	Maximum feasible participation; sunshine laws; deliberative forums	Deliberation; consensus; majority rule
	Courts	Vertical legal authority over society	Laws; statutes; contracts	Precedence; reasonableness; due process; substantive rights
MARKET	Shareholders/owners Consumers	Vertical over management/labor Horizontal with owners	Profit Consumer law	Efficiency Affordability; quality; satisfaction
ADMINISTRATIVE	Principals; supervisors; bosses	Vertical over agents; subordinates; contractees	Performance measures; administrative procedures; organizational charts	Deference to positional authority; unity of command; span of control
	Partners; peers	Horizontal with peers	Written agreements; decision-making crocedures; negotiation regimes	Trust; reciprocity; durability of relationships
	Peers	Horizontal within profession	Codes of ethics; licensure; performance standards	Professional norms; expertise; competence

administrative accountabilities, or intra-administrative accountabilities, such as those found in trade-offs between bureaucratic-collaborative accountabilities (Koliba *et al.* 2011).

In the context of climate change mitigation, a governance network's capacity to support or hinder the democratic accountability of its actors hinges on its capacity to be described as "democratically anchored." Sørensen and Torfing (2005: 201) assert that "governance networks are democratically anchored to the extent that they are properly linked to different political constituencies and to a relevant set of democratic norms that are part of the democratic ethos of society." Democratic anchorage is one of the central governance features of a governance network. However, it has been noted that governance networks that exist at the international scale are confronted with "democratic deficit" because there are no widely accepted, enforceable international democratic norms (Haas 2004). As we consider climate change, international governance needs to be addressed in the light of the network structures that are implicated in certain kinds of climate change mitigation initiatives and the roles that vertical, horizontal, and diagonal relations play in relation to the leadership structure and flow of power and authority. Governance thus needs to be understood in the context of the accountability frameworks that persist within each node (or network actor) as well as across the ties forged between accountees and accounters across governance networks. Comparative analysis of different governance networks, especially analysis of their implicit and explicit performance measures and accountability ties, could potentially inform the design of complex policy regimes dealing with transnational GHG pollution control and other global public policy problems. Institutionalization of accountability through specific policy designs could be informed by such comparative analyses.

The processes of institutionalizing accountability in governance networks merit special consideration as they explicitly deal with the problem of ensuring procedural fairness in complex situations involving a myriad of private and public sector actors. Zia and Koliba (2011) argue, as also emphasized by Bäckstrand (2008), that three criteria of accountability process need to be explored: (1) transparency and public provision of information by a governance network is critical for ensuring that accounters are able to access the information in a transparent manner; (2) monitoring mechanisms ensure whether the governance network has institutionalized monitoring of its stated goals and actions taken to meet those goals; and (3) representation of stakeholders' concerns whether partnerships include government, market or civil society actors. These factors are predicated on the extent to which wide ranging stakeholder groups participate formally in the network, either as lead or as participating partners (Bäckstrand 2008: 82). Zia and Koliba (2011) argue that the public, private or non-profit sector characteristics of actors will matter.

Democratic accountability is rendered when elected officials, citizens, courts, and interest groups are engaged as stakeholders in a transparent manner with monitoring mechanisms that are trusted by all engaged actors. At the international scale, this calls for the reduction of the "democracy-deficit" to enable accountability processes in global climate governance networks. Bäckstrand (2008: 98) presents a comparative analysis of process accountability features for a variety of public–public and public–private climate governance networks. From a governance network accountability framework perspective, these three process criteria—transparency, monitoring and stakeholder representation—may be used to describe the processes activities that are adopted to maintain effective accountability ties.

Within effective accountability ties, performance measures are used to ascertain the extent to which explicit standards, such as performance inputs, outputs and outcomes are being met. The definition of what constitutes effective performance measures for a governance network is a critical question to be addressed. There have been some studies conducted that look at the efficacy of network structures in achieving ascribed performance outputs and outcomes (see as a representative: Marsh and Rhodes 1992; Heinrich and Lynn 2000; Koontz 2004; Mingus 2004; Frederickson and Frederickson 2006; Kelman 2006; Rodríguez et al. 2007; Vining et al. 2005). The highly contextual nature of the environments that governance networks operate within, coupled with the highly contextual nature of most of the perceptions of the network actors within the network, render the development of consensus around common definitions of viable network performance measures very difficult to achieve.

This becomes an even more complex problem when performance measures across governance networks are compared and assessed for their accountability regimes. In environmental governance arenas, generally, it could be hypothesized that governance networks dominated by high greenhouse gas emission countries (i.e., United States and China) endeavor to choose performance measures that maintain the status quo (i.e., minimal greenhouse gas abatement). On the other hand, if governance networks give voice to those countries that tend to be victims of the environmental crisis (i.e., African countries and island nations), the victims tend to choose performance measures that engender maximum feasible change from the status quo (i.e., maximal pollution abatement). Within public–public partnerships, the dominant accountability ties are the elected officials who are responsible for designing and implementing international treaties and protocols. These public–public partnerships are, at least in theory, high in the representation of elected official and citizen interests. Different governance networks are configured with various combinations of high greenhouse gas emitters and those most vulnerable to climate change, for which reason, Zia and Koliba (2011) postulate that intense conflicts over the choice of performance measures are observed. These performance measures are adaptively updated as performance information flows across governance networks increase, and the science governing the environmental problem matures. Zia and Koliba (2011) apply this theoretical accountability framework to a stratified sample of public–public, public–private and private–private climate governance networks and focus the empirical comparison on their choice of specific performance "outcome" measures vis-à-vis accountability ties of the actors in these different governance networks.

Comparative analysis of accountability ties and performance outcome measures across climate change governance networks

The United Nations Framework Convention on Climate Change (UNFCCC) has driven international process to address climate change mitigation and adaptation at the global scale by relying on the voluntary participation of representative country governments. The UNFCCC process represents an example of "public–public" governance network. Under the UNFCCC process, the Kyoto Protocol was a first significant step in setting up a global governance regime for reducing GHGs.

From a governance network analysis perspective, there are numerous other public–public, public–private and private–private governance networks that are simultaneously trying to address climate change mitigation and adaptation issues. Some other examples of public–public climate change governance networks include the Asia-Pacific Climate Change Partnership (APP), International Partnership for the Hydrogen Economy (IPHE), Carbon Sequestration Leadership Forum (CSLF), Cities for Climate Protection (CCP) and Clinton Climate Initiative (CCI). Similarly, some "private–private" governance networks addressing climate change include International Climate Change Partnership (ICCP), World Business Council for Sustainable Development Climate Partnerships, Combat Climate Change (3C) and Greenhouse Gas Protocol (WRI and WBCSD). Finally, some examples of "public–private" climate change governance networks include Renewable Energy Policy Network for 21st Century (REN21), Renewable Energy and Energy Efficiency Partnership (REEP), Joint Implementation projects (JI) under Kyoto Protocol, Clean Development Mechanism (CDM) projects under Kyoto Protocol, World Bank Prototype Carbon Fund (PCF) projects and the United States Environmental Protection Agency's Methane to Markets (M2M) projects. In these examples, regulations, grants and contracts give structure to networks organized through inter-organizational projects or programs.

The governance network accountability framework provides a coherent theoretical tool to compare the design of accountability and performance management systems across public–public, private–private and public–private climate change governance networks. More specifically, Zia and Koliba (2011) apply this integrative framework to a sub-sample of two governance networks for each of the three governance network types: private–private; public–public and public–private, as shown in Table 7.2. They coded performance outcome measures from the documents released by these different types of climate governance networks. The third column in Table 7.2 shows a summary of coded "Performance Outcome and Activity Measures" for the sampled governance networks and their temporal deadlines specified by these governance networks.

Zia and Koliba (2011) then undertook comparative interpretive analysis (Yanow 1999) of these three governance network types by specific performance outcome and activity measures agreed upon by the sampled governance networks. From this comparative interpretive analysis, they derived four

Table 7.2 Performance measures (activities and expected outcomes) and their deadlines across sampled climate change governance networks

Type of governance network	Sampled climate governance network	Performance measures	
		Activities and expected outcomes	Deadlines
Private–Private	ICCP	Address continued growth of greenhouse gas emissions through mechanisms such as emissions trading. Business and industry expertise are important parts of this process. Technological innovation is crucial.	Vague
	3C	Businesses cooperate to reduce emissions for a stable climate by putting a price on carbon emissions, setting minimum efficiency standards, encouraging sustainable forestry and agriculture, and pushing low carbon technologies.	Vague
Public–Public	UNFCCC	Countries coming together to consider what can be done to reduce global warming and to cope with whatever temperature increases are inevitable. The Kyoto Protocol sets binding targets for thirty-seven industrialized countries and the European community for reducing greenhouse gas emissions by an average of 5 percent against 1990 levels over a five-year period. Kyoto mechanisms include emissions trading, Clean Development Mechanism (CDM) and Joint Implementation (JI).	Reductions must be met over the five-year period 2008–2012.
	APP	Overall goal is to accelerate the development and deployment of clean energy technologies. There are sub-goals regarding energy security, national air pollution reduction, and climate change. The partnership will focus on expanding investment and trade in cleaner energy technologies, goods and services in key market sectors.	Vague

		Project	Description	Crediting period
Public–Private	CDM	Yiyang Xiushan Hydropower Project, P.R. China	Reduce CO_2 emissions by 243,043 metric tons per year by using a consolidated methodology for grid-connected electricity generation from renewable sources.	Crediting period of 5/10/9–5/9/16 with lifetime of project lasting thirty-three years from 8/18/5
		Casa Armando Guillermo Prieto – wastewater treatment facility for a Mezcal distillery	Reduce CO_2 emissions by 15,153 metric tons per year by using thermal energy with or without electricity and methane recovery in wastewater treatment.	Crediting period of 5/7/9–5/6/16 with lifetime of project lasting twenty-five years from 4/23/07
		Heilongjiang Chemical N2O Abatement Project	Reduce CO_2 emissions by 279,319 metric tons per year by implementing catalytic reduction of N2O inside the ammonia burner of nitric acid plants.	Crediting period of 5/7/9–506/16 with lifetime of project lasting twenty-one years from 7/17/7
	JI	Timisoara Combined Heat and Power Rehabilitation for CET Sud location	Upgrade the existing heat production plant CET Timisoara Sud with cogeneration capacity.	Project lifetime is twenty years as of September 2005
		Debrecen landfill gas mitigation project	Installation and operation of a new landfill gas collection system. Reduction in CO_2 emissions by 413,866 metric tons over crediting period.	Crediting period of 1/1/8–12/31/12, with lifetime of project lasting ten years from 11/30/7
		Revamping and modernization of the Alchevsk Steel Mill	Replacement of technology and upgrade of all major components of iron and steel making and finishes processes.	Crediting period of 1/1/8–12/31/12, with lifetime of project lasting forty years from 8/24/5

performance management dilemmas that currently bedevil the institutionaliza-tion of accountability in global climate governance. They call these dilemmas of strategy, uncertain science, integrating multiple scales and verification, each of which is described below along with the findings of comparative interpretive analysis on the performance outcome measures across these governance net-works shown in Table 7.2. Zia and Koliba (2011) postulate that a meta-level res-olution of these dilemmas is critical for institutionalizing accountability in global climate governance; however, there are no global level institutions in place to enable this kind of meta-level resolution across governance networks. The impli-cations of these dilemmas on post-Kyoto international climate policy design are discussed on page 155.

Dilemma of strategy: How to develop consistent performance measures
when different governance networks propose different measures, such
as GHG/year, GHG/BTUs and GHG/capita?

At the international scale, each GHG polluting nation is caught up in proposing a set of performance measures that, by definition, either let that nation free ride or incur minimal abatement costs. Under the UNFCCC negotiated Kyoto Proto-col, which represents a public–public type of international governance network dominated by GHG high emission countries, "grandfathering" performance measures were adopted despite calls for GHG/capita based performance meas-ures by developing countries who are expected to bear the most adverse impacts of climate change (IPCC 2007). The UNFCCC based public–public governance network was thus co-opted by the strategic goals of rich-developed countries into adopting a grandfathered performance measure (reduce GHG/year emis-sions by a target year below certain baseline year). Interpretive analysis of recent Conference of Parties (COP 15) negotiations in Copenhagen and COP 16 nego-tiations in Cancun for a post-Kyoto UNFCCC based international treaty shows that grandfathering based performance goals are also being considered for a post-Kyoto Treaty.

As shown in Table 7.2, for public–public climate change governance network of UNFCCC, performance outcome measure of reducing GHG/year by ~5 percent below 1990 level by 2008–2012 was set as a binding commitment for Annex I parties who ratified the Protocol. This performance measure is an example of "grandfathering," which has been compared in the literature with some other performance outcome measures, such as GHG/capita that was not adopted by the UNFCCC governance network (Najam and Sagar 1998; Bier-mann 2005). The choice of performance outcome measures within this public–public governance network is thus fraught with political maneuvering and strategizing by network actors, as discussed in the previous chapters. This is in the interest of rich industrialized countries, who happen to be the major GHG polluters as well, to choose a performance outcome measure that by definition minimizes their GHG emission reduction burdens. Grandfathered targets agreed upon in the Kyoto Protocol, as compared to the GHG/capita type of performance measures, apparently do exactly what serves the interest of rich industrialized countries. The accountability analysis of the UNFCCC governance network thus shows that the choice of performance outcome measure is an artifact of political power and scientific knowledge, which overrides ethical concerns of equity raised by developing countries who have consistently argued that GHG/capita performance measure must be chosen by UNFCCC (Najam *et al.* 2003; Pettenger 2007; Cass 2006).

In contrast to UNFCCC governance network, the APP governance network has remained vague in setting any performance outcome measures, as shown in Table 7.2. The APP in fact argued that there should be no binding performance measures, which again demonstrates the tragedy of commons as APP represents the most sizeable GHG polluting countries. Similar vagueness is obvious from the performance measures developed by private–private climate change governance networks—ICCP and 3C—shown in Table 7.2. Under public–private partnerships of CDM and JI case study projects, there are specific performance outcome measures that have typically very long target dates (as shown in Table 7.2).

Comparative application of the accountability framework across public–public, private–private and public–private type of climate governance networks reveals variegated patterns of performance outcome measure selection that is contingent upon the type of actor configuration in a particular network. Zia and Koliba (2011) call it a "dilemma of strategy" in setting up performance standards in complex governance networks. This dilemma is, for example, obvious when one considers APP governance network. After the Bush administration in the United States reneged on the U.S. commitment to sign the Kyoto Protocol on the pretext that developing countries were not included, the U.S. government, in alliance with other countries that consider UNFCC process as too burdensome and potentially a costlier enterprise, decided to engineer a governance network of seven highest GHG polluting countries that they call APP. These seven countries are responsible for at least 50 percent of the current global GHG emissions. The performance outcome measure that APP proposes is no binding commitments to reduce GHG emissions. So, APP does not want a performance standard at all. When criticized for this, some APP leaders called for GHG/BTU and BTU/GDP (i.e., intensity-based) performance standards, which are practically business-as-usual scenarios of growing GHG emissions in the atmosphere.

Dilemma of strategy thus demonstrates that different governance networks, based upon the differential goals and accountability frames of accountees and accounters in the governance networks, propose performance standards in tragedy of commons situations that minimize actor level cost of pollution abatement. When there are multiple governance networks in public, private and public–private domains with variegated performance outcome measures, it becomes very difficult to hold any governance network accountable on a common performance outcome measure because they do not agree with a common performance outcome measure to begin with. A more serious and intractable horn of the dilemma concerns the fact that the accounters for multi-actor configurations in different governance networks are not interested in holding network actors responsible on some unified performance outcome measures due to the inherent nature of their value and goal conflicts, or in the case of future generations, mere absence of actors. Furthermore, the inherent trade-offs that persist in complex governance networks may be viewed in terms of competing perspectives from the elected officials of country governments in public–public governance networks or as trade-offs between democratic accountabilities and market accountabilities in public–private partnerships.

Dilemma of uncertain science: How to incorporate scientific uncertainty in policy design?

In this unfolding tragedy of commons, actors across various climate governance networks have strategically deployed scientific uncertainty to their advantage. In the UNFCCC governance network, for example, the controversy of whether to consider existing forests as carbon stocks or not, and by how much, provides an interesting case-study of this dilemma (e.g., Hirsch *et al.* 2011). While there is large scientific uncertainty about the carbon uptake functions of forest systems in evolving climatic conditions, some network actors with large standing forests argued for inclusion of forests as carbon sinks. However, other network actors argued against the inclusion of forests, citing scientific research showing diminishing carbon uptake in higher CO_2 concentrations. Inclusion or exclusion of forests as carbon sinks presents one example of dilemma of uncertain science, as it might be too late to take policy action for or against deforestation by the time scientific uncertainty is reduced.

Another example of this dilemma concerns the differential weights that are accorded to different GHGs based on their CO_2 equivalency. While UNFCCC aimed at standardizing these weights, there has been a severe critique of the methods used to standardize the weights (IPCC 2007). Some private–private governance networks have expressed their concerns that industrial gasses are accorded much higher weights, while some other governance networks have argued the opposite, i.e., the industrial gases should have been accorded even higher weights due to their higher radiation potential. Additional questions about "latent" GHG emissions and their inclusion in UNFCCC basket of post-Kyoto gasses remain largely unaddressed, as well.

The scientific uncertainty about climatic change impacts and how it translates into different positions, especially trade-offs between mitigation and adaptation, pose another set of problems in setting up accountability mechanisms. For some governance networks, increased investments in adaptation strategies will entail higher benefits for future generations of citizens (accounters in the accountability model). Other networks argue for higher investments in mitigations strategies. The lack of scientific certainty about the nature and extent of climate impacts poses daunting challenges for designing efficient and fair policies at multiple generational time-scales.

From the comparative perspective, the dilemma of uncertain science may be viewed in terms of trade-offs between professional accountability and either elected official or market accountabilities across various types of governance networks. A country's failure to take climate change seriously may be fueled by allusions to a scientific uncertainty that is being tied to climate change models. In the midst of this scientific uncertainty, debates over how to value carbon sinks may be exploited by certain stakeholders as a lack of professional consensus from the scientific community.

Dilemma of integrating multiple scales: How to integrate emission entitlements across multiple space-time scales?

Climate change mitigation actions are being proposed at multiple space-time scales by different governance networks, which imply that the accountability challenges of measuring their respective performances also multiply with multi-scalar mitigation actions. Double, or even triple, accounting of the same "mitigation action" is the biggest concern here. Consider the example of a wind turbine installed in a small town in Europe, for which a city in the CCP governance network claims credit, a firm in ICCP claims credit, and a country in UNFCCC claims credit. In fact, in some voluntary air travel GHG emission offset systems, gross instances of double or triple accounting have been reported for the same set of carbon sinks that are used as GHG emission offsets.

Resolving this dilemma at the inter-governance network level will pose a huge challenge as each governance network and its respective actors have the incentives to undertake double or triple accounting. There has been some movement towards unifying these cross-scalar mitigation activities in terms of a consistent scale, but this remains a huge challenge on many fronts. Consider the example of a huge multinational corporation operating in many countries. Should their mitigation actions in countries of their operation be ascribed to host countries or the country of their headquarters? Given the typical accounter goals of profit maximization in private–private climate governance networks, Zia and Koliba (2011) postulate that public–public or public–private partnerships might be more effective in reducing multiple accounting of the same emission reduction credits. In purely private–private partnerships, there essentially is no democratic accountability and very little, if any, market accountability driving voluntary compliance of mutually determined perform-ance measures.

Different governance networks operate at different geographic and social scales, as shown in Table 7.2. The question of multi-scalar accountability may be viewed as trade-offs between accounters and accountees at these different geographic and social scales. The public–private partnerships developed for CDM and JI projects operate in specific geographical conditions and temporal scales that are different from the performance outcome measures agreed upon in public–public and private–private type of governance networks. Integration of performance outcome measures across these multiple space-time scales is, perhaps, impossible, while issues of double or triple accounting pose daunting challenges for comparing "observed" performance outcomes claimed by different governance networks.

Dilemma of monitoring and verification: How to monitor and verify performance benchmarks?

Monitoring and verification of claimed mitigation actions poses another set of challenges. There has been a movement towards third party verification of emission reductions that get claimed (e.g., growing California Climate Action Registry Contracts). In a third party verification system, accounters hire an independent third party to verify whether accountees have actually reduced the claimed emissions. Despite the proliferation of third party verification systems, there are some monitoring and verification issues that cannot be easily resolved. Consider the example of CDM public–private partnerships established under flexibility mechanisms of Kyoto Protocol. There is no consensus about how to establish baseline "reforestation" or "afforestation" scenarios in developing countries that are eligible to claim CDM based emission reduction credits because they are so dependent upon how one calculates baseline scenarios. Some critics argue that CDM has provided perverse incentives to many developing countries to enhance their GHG emission rates so that they could receive more GHG emission reduction credits when lower emission rates (as opposed to exaggerated baseline rates) are verified. Similar challenges exist for the REDD+ (Reduced Emissions from Degradation and Deforestation) policy mechanism.

Verification of some GHGs is relatively easy (e.g., some industrial gasses), while other GHGs pose persistent dilemmas. Point sources of GHG emissions (e.g., industries) can be easily tracked, but non-point sources (e.g., transportation systems) are not easily amenable to verification. Accurate measurement of transportation activities and transportation behaviors poses age-old modelling dilemmas. The variance of estimates tends to be high. There are also strategic problems with respect to some transportation activities, e.g., military-based transportation operations are typically not reported. Accurate quantities of energy consumed by military activities are not verifiable due to strategic security problems with revealing the nature and extent of these activities. Overall, the governance networks need to develop the capacity to become more effective in verification processes, especially third party certifications. However, as recently evidenced during Copenhagen negotiations of COP 15, GHG polluting countries such as China refused to institutionalize third party verification mechanisms because they considered these independent verification measures as "infringements on their sovereignty."

The dilemma of verification is fueled by some of the same trade-offs found in the dilemma of uncertain science. Gaps in the scientific models may be exploited by detractors of climate change, thereby undermining the authority of professional accountability. Scientific verification is also confounded by a range of administrative burdens that accompany verification processes. Which actors have the administrative authority to collect, analyze and verify data? To what extent should self-reported data be accepted? These challenges suggest that the administrative lines of accountability (both bureaucratic and collaborative) are hard to clarify and put into practice. These questions speak to the authenticity of administrative accountabilities that exist in climate change mitigation governance networks.

Accountability framework and post-Kyoto climate governance regime

Climate change mitigation strategies are perceived to be undertaken by a large variety of governance networks that present particular accountability challenges. The inherent complexity of climate change governance is fueled by a range of perverse incentives that lead to global "tragedy of commons" for economically vulnerable actors as well as future generations. Zia and Koliba's (2011) comparative analysis reveals that less transparent processes, ineffective monitoring mechanisms, inadequate stakeholder representation and lack of consistent performance outcome measures across governance networks gives rise to at least four chronic dilemmas that require meta-level resolutions for institutionalizing inter-governance network level accountability mechanisms. They call these the dilemmas of strategy, uncertain science, multiple scale integration and verification.

If the post-Kyoto climate governance regime that is now being negotiated across a range of governance networks attains the same performance levels as prior efforts, human civilization is very likely to initiate a dangerous spiral of positive feedback loops of GHGs under business-as-usual scenarios. The resulting cascading effects will be difficult to reverse due to atmospheric complexity and non-linear lagged effects (IPCC 2007). More recent climate science, since the fourth assessment IPCC report, presents even grimmer picture (Raupach *et al.* 2007; The Copenhagen Diagnosis 2009). It is critical that a post-Kyoto climate governance regime incorporates accountability-driven design features that ensure that anthropogenic GHGs stay well within planetary resilience, prior to the initiation of dangerous positive feedback loops.

If humanity remains trapped in these dilemmas, worst-case climate change scenarios are very likely to materialize. This trap need not be inevitable. The climate change governance networks, at both the political and strategic levels, could design governance networks by drawing on the systematic accountability frameworks presented here. Transparent processes need to be promoted to enable cooperative resolutions of these dilemmas. However, this will require meta-level comparative policy analytical thinking and political resolution. The reduction of "democracy-deficit" in international/global governance networks could be the first step in this journey. The idea of democratically elected global parliament could reduce global democracy-deficit (Biermann *et al.* 2012). Acknowledgement of inter-generational accountability issues could be another step. The challenges of asymmetric power and knowledge distribution among the actors in governance networks will nevertheless continue to bedevil meta-level political efforts aimed at resolving these dilemmas. More comparative policy analytical research is needed to understand how the feedback loops of institutionalized accountability mechanisms across climate change governance networks affects the emergence of power and knowledge distribution asymmetries at the global scale. Understanding global climate change policy design problems like those discussed here can inform the development of a coherent accountability framework that simultaneously takes into account actors, processes and outcomes.

Adaptation Fund

The Adaptation Fund (AF) was established under the Kyoto Protocol as a financial instrument to support projects in developing nations vulnerable to the effects of climate change. It is expected to be continued under a post-Kyoto governance regime, albeit its name may be changed to "Green Climate Fund." It is a significant policy that needs to be developed given the accountability of industrialized countries in their contribution to greenhouse gas emissions with respect to the effects of climate change and developing countries' limited capacity for coping.

Currently, the AF is funded by a 2 percent levy on certified emissions reductions (CERs), colloquially known as carbon credits, issued by qualified initiatives under the Clean Development Mechanism. While the exact cost of adapting to a changed climate cannot be known, the AF's projected maximum payment of $438 million by 2012 falls shy of the estimated cost of $500 billion per year over the next fifty years, with many of the highest costs expecting to be incurred in world's least developed nations (Parry *et al.* 2009). In addition to an inadequate amount of funding, the AF's funding scheme works to the detriment of the Clean Development Fund by taxing developed countries seeking to invest in clean technologies abroad. Therefore, it has been proposed frequently that the Adaptation Fund change its funding mechanism. Further, the AF also needs to establish a clear criterion for eligible adaptation projects in order to adhere to its mission and promote resiliency in economically disadvantaged areas.

Whereas distributive justice has been an issue of contingency for CDM and JI policy mechanisms, it is a key concern with regards to the Adaptation Fund. The Fund, which is funded by 2 percent of CERs generated, as well as through private funds and donations, is specifically designed to provide the most vulnerable countries with the least adaptive capacity with compensation for the pollution that industrialized countries have caused. Grasso's work (2011) traces the meetings that gave birth to the Adaptation Fund as it exists today and supports that through the clashes and negotiations and revisiting of its terms, developed and developing countries have been able to equally and actively participate in its planning. As the Fund is designed to be more accessible to the least developed countries, they have distributed the power more greatly to and evenly among national governing bodies rather than multilateral agencies. In the case of the AF, negotiations continued until developing countries were satisfied that the Fund ensured that the most vulnerable countries received a fair disbursement of funds and the management of funds would be undertaken by entities less likely to favor the developed countries.

While Grasso's assessment of the procedural ethics of designing the AF is quite positive, the operationalization of it has not seen the level of success in disbursing funds as was planned. While the Fund's interdependence from official government assistance was planned to make funding more equitable, Horstmann (2011) explains that besides the two percent of CERs generated through the CDM, there were no binding financial agreements to create funding. Ciplet *et al.* (2011) agree that with the vagueness of funding types/sources, no one is required to take responsibility to see that there is adequate funding. The concept of "direct access" to the Fund for vulnerable countries has rather evolved into a set of hurdles to jump through in order to comply with the policies and procedures required to access the funds from a Global Environmental Facility (GEF) agency, although the GEF's role was diminished for that very reason. Horstmann also questions the lack of specificity of the initiative to particularly support the most vulnerable countries. Relevant documents regarding the AF mention the prioritization of "vulnerable countries," but merely in generalizations to their geographic vulnerability to climate change or the economic vulnerability to adapt themselves, and without specific language regarding who these priority countries are and whether funding needs to be provided to specific vulnerable communities, state governments, specific projects, etc. Also least developed countries that are not party to the Kyoto Protocol cannot be considered. The National Implementation Entities in charge of the AF now face the challenge of having the capacity to consider and assess each vulnerable country through a transparent process that provides adaptation funding as the AF promises.

In order to receive funding, the AF depends upon the conscience of industrialized countries to hold themselves responsible for past, current, and future climate change due to their pollution. Grasso points out that it is not meant to be a charity fund, but deserved compensation; unfortunately, that ideology is not shared or cared about enough amongst many of the worst GHG polluters. Also, many countries feel that their debt has been paid through other initiatives and do not acknowledge the different form of capacity and targeted approach that the AF can provide. Ciplet *et al.* (2011) urge for a mechanism for countries to exemplify their acceptance of their burden upon the environment through some form of calculation of past pollution and potential harm and how they can fairly account for that.

At present, the AF has vague standards for evaluating potential adaptation projects. Eligible recipients need only to qualify as "developing country Parties to the Kyoto Protocol that are particularly vulnerable to the adverse effects of climate change in meeting the costs of adaptation" (Hosrtmann 2011: 1091). The danger of such nebulous language is that it is highly subjective to interpretation. As a general principle, the Adaptation Fund Board will award funds to a project that provides "economic, social and environmental benefits to the most vulnerable communities" (Horstman 2011: 1093). Many of the projects currently underway emphasize the development of water quality and agricultural production.

Because the AF sources its funds from a tax on CERs, it essentially discourage carbon offsets. While this tax is not particularly significant detrimental at the

2 percent level, the previously demonstrated need for significantly greater funding necessitates a source that will not be detrimental to the Clean Development Mechanism. Recalling that the Adaptation Fund is an attempt to reconcile the inequity caused by the pollution of developing countries, it would seem prudent to adopt a transportation tax scheme. Flåm and Skjærseth (2009: 112) argue that a tax on maritime and air fuel provide the benefit of mitigating usage, thus reducing environmental impact, as well as raising revenue. A fuel tax would also not be as susceptible to repeal as a carbon tax, they argue. Furthermore, the tax could be levied without a significant burden on the individual consumer, for example, a $10 dollar tax per flight could yield an estimated $20 billion per year (Flåm and Skjærseth 2009: 112).

Zadek (2011: 1065) proposes that "[i]t is national (and, perhaps in some cases, sub-national and regional) leadership and ambition that will drive progress." Zadek (2011) does acknowledge that outside funding would be helpful, but is mostly a proponent for moving away from placing blame and more looking towards how to fix the problem. However, the reality of the situation is that climate change and greenhouse gas emissions are a global problem. There are also portions of the global community who have contributed huge amounts more to this problem for which everyone has to suffer the consequences. Because of this, it is important that regions of the world such as China and the United States take on the larger consequences of their actions. With this methodology and mentality in place, there are ways in which the industrialized nations could and even should help to aid the underdeveloped countries to prepare for adaptation to climate change.

The Adaptation Fund is managed by the Adaptation Fund Board (UNFCCC). The Adaptation Fund Board consists of representatives from 16 countries that adhere to the Kyoto Protocol. Climate adaptation in a country or a community is not isolated; it is connected with all other social, environmental, and economic problems in that area. Climate change will affect all aspects of life in vulnerable areas including: food availability, poverty, unemployment, access to water, sanitation education, and healthcare (Ziervogel and Taylor 2008). That is why it is so important for local stakeholders to have input and decision-making power regarding what they need, and what solutions would work best in their community. The AF board has little contact with members of civil society or stakeholders; the only contact is the allowance of accredited observers and possible formal dialogue with those observers at the end of the board meetings (Abbott and Gartner 2011).

The members of the board and distribution of decision-making power demonstrates the "state centric" approach of the AF. The fund is designed based upon the involved states desires; the desire of industrialized countries for accountability for results, and the desire of developing countries for adequate resources that they can allocate themselves (Abbott and Gartner 2011). These are concerns at the national level that do not address the concerns of states, local communities, or civil society. A study by Anna Taylor and Gina Ziervogel (Ziervogel and Taylor 2008), regarding priorities of stakeholders at the local and municipal level in South Africa, found that "[t]he lack of effective communication between stakeholder groups (villagers, government officials, and researchers) is curtailing participation in decision-making and thereby disempowering all from facilitating positive change (in the form of adaptation and sustainable development)." The lack of communication between stakeholders and the members of the Adaptation Fund Board, who control their funding and therefor the focus of their projects, could impede on the success of their climate adaptation projects.

The Adaptation Fund was "established to finance concrete adaptation projects and programs in developing country Parties to the Kyoto Protocol that are particularly vulnerable to the adverse effects of climate change" according to the United Nations Framework Convention on Climate Change. Despite this goal of supporting "particularly vulnerable" communities there is no official definition of "vulnerability" with which to rate countries applying for aid (Horstmann 2011). When applying for funding each country has to make a case for their vulnerability, so they are forced to define it themselves. This system again demonstrates the country driven approach of the AF. The allocation of funding to the most vulnerable communities is decided by the national government of each country and therefor responsibility and accountability for the success of the projects is placed not with the Adaptation Fund Board but with each country's national government.

One of the national concerns that has been voiced repeatedly is the difficulty in accessing climate financing (MacLellan, 2011), especially for small countries like the Pacific Islands that do not have the capacity to manage increased flows of funding. This demonstrates the lack of an appropriate infrastructure; receiving the funding is only one of the first steps in this adaptive process. Each country needs to increase their adaptive capacity by creating an appropriate infrastructure designed for that country addressing what it needs to accept and manage increased funding, and have an educated staff and that can gather, prioritize, and understand the needs of the country across all scales, and finally implement these changes. This infrastructure has the potential to increase the efficiency of many donor-funded projects concerning various socio-economic concerns. This is important because funding tends to shift with environmental and social fads (Ziervogel and Taylor 2008).

Over the years the international community has been slow to identify and articulate concrete financial adaptation mechanism needed to help developing countries meet the threat of climate change. This is exacerbated by the fact that the need for adaptation is very much dependent on how much the international community is able to mitigate. The less we decrease our carbon emissions the greater the need for adaptation will be. One of the problems with adaptation is that it's hard to measure the cost. This is largely due to fact that the need for adaptation funds is directly linked with other development issues such as population growth, economic development, and conflicts.

One of the main threats to the Adaptation Fund is its legitimacy. Negotiations at the Marrakesh and Bali summits in 2001 led to disagreements between "developed countries" who "thought it was only natural that the GEF (Global

Environment Facility) would be operating the AF in a manner similar to the other funds" and developing countries. The end result was a win for developing countries who succeeded in establishing a "GEF-independent board with major-ity representation of developing countries" and "they managed to 'reduce' the GEF and the World Bank's roles" in the fund (Flåm and Skjærseth 2009). The Adaptation Fund is an important program in order to reduce the effect of climate change on poor countries. Both rich countries and poor countries agree that some exchange of wealth will have to be part of the solution.

The AF makes sure that the developing countries do get the representation they need. The AF board is made up of 16 seats and 16 alternate seats, and the majority of the seats are held by developing countries. Special seats have also been made for the least developed countries and the small island developing states (Rübbelke 2011).

Klein *et al.* (2005) claim that they are uncertain about the efficiency and effectiveness of AF-like programs. They wonder whether AF will lead to devel-oping countries relying on these projects in order to deal with climate change. If people start to rely on adaptation alone, then climate change will reach a level where the only effective adaptation will come at a very high economic and social cost. Economic cost is their last issue with AF. Currently, contributions to AF are voluntary, so they are not meeting the goal that will allow them to complete all these projects (Klein *et al.* 2005). Also questioning economic cost, Tan (2008) wonders whether the initiative led by the G8 (group of eight large world economic powers: Canada, France, Germany, Italy, Japan, Russia, United Kingdom, and United States) will undermine existing negotiations surrounding climate change. In addition, they worry that the rush to fund AF projects will lead to an establishment of top-down funding, which could lead to a lack of par-ticipation from developing countries, and will not be able to reach sustainable transformations (Tan 2008).

Many developing countries, however, currently do not even have the relevant "absorptive capacity"—the capacity to carry out the adaptation measures needed—even if the funding were available. Most will unnecessarily have to suffer adverse impacts of climate change that could be avoided under an improved adaptation regime. The responsibility for these avoidable adverse impacts—whether due to a lack of funding or of absorptive capacity—will fall squarely on industrialized countries. Some stakeholders, mainly from the devel-oped world, have been tempted to cite the lack of certainty about the adaptation funding needs of developing countries and their lack of absorptive capacity as reasons to postpone a debate of the rough issue of international adaptation finance. The two issues are intricately linked, and there is an urgent need to look into ways of simultaneously scaling up the provision of adaptation funds for developing countries of the appropriate kind, and the absorptive capacity to use these funds meaningfully (Flåm and Skjærseth 2009).

Presently, all international adaptation funding instruments—except for the Adaptation Fund—are replenished through official development assistance (ODA)-type bilateral donations. The level of international funding for adaptation in developing countries is inadequate to meet projected needs. The current bilateral donation instruments are unlikely to ever be able to generate the required levels of funding. Moreover, adaptation funding is seen by most developing countries not as a matter of "donations" but as one of costs imposed by developed countries, and as such as debt incurred by them. Therefore, the traditional ODA funding modes—grants or concessionary loans—are seen to be inappropriate payment modes. Further innovative financing mechanisms apart from the CDM Adaptation Levy are needed to fill the adaptation "funding chasm" (Flåm and Skjærseth 2009).

Internationally, for the fund to be successful, funds for adaptation need to be allocated on a strategic basis and not involve international micro-management at the project level. This strategic allocation of international adaptation funds should use the existing international bodies and initiatives to allocate funding streams, and not try to duplicate them under a "climate change banner" (Flåm and Skjærseth 2009). Domestically, as mentioned above, there is a need to enhance "absorptive capacity" not only at the project level, but also at the level of domestic policy as well. Perhaps the most unique and exciting aspect of the Adaptation Fund is its governance. The governance of the Adaptation Fund represents a milestone in the evolution of international funding mechanisms, since for the first time developing countries have genuine ownership of such an instrument (Flåm and Skjærseth 2009). In the case of adaptation funding, developing country ownership and public transparency of decision-making is not only desirable but also a prerequisite for success, particularly in the context of mainstreaming activities. Given this, the Adaptation Fund should be the main instrument for the purpose of raising and managing of international adaptation finance for developing countries and should absolutely be included in a post-Kyoto governance regime.

8 Governing environmental complexity

I have argued that the phase spaces of coupled human and natural systems could not be pre-defined to "scientifically manage" human-induced global climate change. Instead, I have argued that a complex systems based adaptive, decentralized and democratically anchored governance of coupled human and natural systems could be used as guideposts to adequately cope with global environmental and social crises. Normative underpinnings of governing complexity, informed by complex systems based understanding of global climate change, global biodiversity loss, and global food insecurity could guide us towards adaptive governance-based interventions in local to global communities. In turn, this could create room for open-ended and value-laden adaptive governance strategies to cope with global climate change, global food, and biodiversity loss and other such problems. Iterative nature of adaptive risk governance could also provide opportunities to the local managers of complex systems to use creative problem solving skills based upon the community values and the constantly evolving knowledge of the partially unknowable adjacent possible states of coupled human and natural systems that could be dynamically managed with the evolving knowledge of adjacent possible states containing both intended and unintended consequences.

At the global scale, these normative underpinnings for a post-Kyoto global climate governance regime carry important ethical and legal implications. In the midst of unmitigated, or even partially mitigated, human induced global climate change, human civilization is witnessing perhaps the greatest crime that continues to be perpetrated by powerful industrialized nations since 1750 and now closely followed by rapidly industrializing countries through the fossil fuel driven development path discussed earlier in this book. Even if we hit a Hubbard's peak for oil in the twenty-first century, the availability of technologically recoverable coal and natural gas for the next 300 to 500 years will certainly condemn the present human civilization, as we know it, literally and figuratively to the hell. This condemnation will spare neither the rich industrialized countries nor the poor industrializing countries.

While the "Newtonian" system of the world, which brought industrial revolution to the present human civilization, might have engendered an illusion of control by human systems over the natural systems, the sheer complexity of the natural system's response to a quadrupling of CO_2 emissions 100 years from today (2012) is simply unfathomable, much less controllable. The human understanding of climate change impacts under tripling or quadrupling of CO_2 emissions above the pre-industrial levels is rather poor, and perhaps beyond our cognitive reach. The non-linearities in the response of natural systems, which could include widespread floods, droughts, heatwaves, and storms, could play havoc with agricultural and ecological systems that provide constitutive support to life systems on this planet. The critical transitions in the phase spaces of atmospheric system that regulates weather patterns on this planet are very hard to predict, yet these phase transitions will take place if human civilization continues on the business-as-usual path of tripling or even quadrupling CO_2 emissions above the preindustrial levels.

Lack of policy action at the global scale on mitigation and adaptation to global climate change, which cascades down to national and local scales, carries the potential for this planet to move on a trajectory of "post-human world." This trajectory does not have to be "inevitable." The human civilization contains the seeds for intelligence and normative actions, which could trigger ethical, legal and institutional changes away from the business-as-usual development trajectories. Early warning signals through widespread droughts, more intense tropical storms, and unbearable heatwaves could perhaps lead to widespread distress and suffering, which in turn could induce an institutional response to deal with global climate change. Will it be too late then?

Giddens called it Gidden's Paradox. It states that:

> since the dangers posed by global warming aren't tangible, immediate or visible in the course of day-to-day life, however awesome they appear, many will sit on their hands and do nothing of a concrete nature about them. Yet waiting until they become visible and acute before being stirred to action will, by definition, be too late.
>
> (Giddens 2009: 2)

I completely agree with one horn of the dilemma in Gidden's Paradox and this has to do with the lags in the response of global climatic system to doubling, tripling or even quadrupling scenarios of GHGs above the pre-industrial levels. Natural systems are complex systems that exist in alternate stable states. These alternate stable states are "stabilized" by the basins of attraction, which are widely observed in complex systems and the stability induces a certain built-in levels of resiliency in many natural systems. Perturbation of atmospheric system will cause this built-in resiliency to induce a lagged effect before one or many phase transitions occur. So, yes, there is a lagged effect, which is both good and bad news. The good news is that this gives human civilization more time to change its development trajectory. The bad news is that nobody knows how much more time is there, as the planetary system may already be committed to phase transitions, but with an uncertain lagged effect. This sheer uncertainty about the timing of phase transitions, coupled with the lags and inertia in natural systems, poses a significant challenge for human civilization to respond to this crisis at the global scale in a meaningful fashion by weaning off of the fossil fuel development paths.

I do, however, disagree with the second horn of Gidden's paradox that assumes that "the dangers posed by global warming aren't tangible, immediate or visible in the course of day-to-day life." This is a classic example of politics of scale, both spatial and temporal discounting that has been extensively discussed in Chapters 2 and 3. As climatic shifts proceed in non-linear fashions through phase transitions at the planetary scale, it is a huge fallacy to assume that climate change risk is intangible, non-immediate or invisible in the course of day-to-day life. The drought in the Sahel region of Africa, intensified by climate change since early 1970s, has been studied by political scientists, such as Michael Glantz (1977), to understand human response to such crises. Yes, Africa is distant from the United States, more so than the EU! But drought is not going to *only* afflict the Sahel region of Africa under business-as-usual global climate change scenarios. Many parts of the United States are also expected to experience droughts as the time proceeds forward. While current U.S. residents might have forgotten the "dust bowl" era of 1930s, the anticipated droughts in the southwest and perhaps also the southeastern United States will not be far off those of the Sahel region of Africa. These droughts will very likely be "in the neighbourhood." The midwestern United States is called the "bread basket of the world," but a prolonged drought in the midwestern United States could play havoc with the agricultural system, which in turn could also induce widespread migrations. On the other hand, floods are predicted in northeastern and northwestern corners of the United States. Similarly, it will be a fallacy to assume that the residents in the EU will be spared the "immediate" wrath of the global climate change impacts. Given the loss averse and risk averse nature of many humans, as discussed in Chapter 5, the immediacy of such natural hazards induced by climate change perhaps provides the greatest hope against unmitigated global climate change. Ironically, Gidden's Paradox is not inevitable due to this "immediacy" and "para-scalar" potential of global climate change impacts.

This global problem calls for a global scale governance regime that holds high GHG emitters accountable and treats the high emitting nations, firms, and individuals as environmental criminals. Poorly mitigated global climate change is a crime committed by the past and present human generations against the future human and other biological generations. One way to confront the politics of intergenerational scale is to bring the perpetrators of this crime against human civilization to a global parliament or an international court of justice. The irony is that there is no global parliament, while international courts of justice have no teeth to enforce judgments. The leading GHG polluting nations also happen to be the most powerful in terms of their military and geostrategic reach across the globe. The vulnerable of the present generations, and the representatives of future generations who think like a planet, will need to band together and fight against the powerful GHG polluters to enable an institutional change and confront the politics of scale, ideology, and knowledge.

Further, the marketization of climate governance experiment has broadly failed in arresting the growth of GHG emissions. Policy-makers in the international community need to wean away from the marketization approach as embodied in the CDM and REDD+ policy mechanisms. Confronting the politics of knowledge embodied in marketization of environmental goods and services is perhaps the single most important lesson that the human civilization can learn from the Kyoto policy regime. Instead, market mechanisms that are causing the problem of GHG emissions need to be dealt with directly, both from the supply and the demand sides. From the supply side, major sources of GHG pollution are coal, natural gas, and oil industries. The energy and transportation markets, where these energy sources are used as inputs, need to be directly confronted at the international, national and regional/local scales. Introduction of carbon taxes in these markets will be the most efficient policy mechanism. Since carbon taxes can be regressive, some of the revenue from these taxes can be used to provide relief to poor and middle-income groups in industrialized countries. Further, early adoption of carbon taxes in the energy and transportation sectors of the industrializing countries can incentivize the development of energy efficient and renewable energy technologies.

In addition to the fossil fuels, tropical deforestation is another major source of global GHG emissions. Two most effective ways to confront this issue concern the design of international trade mechanisms and shifting the patterns of urbanization in the tropical countries. Current emphasis in the international trade regime established through the WTO body of regulations is based on the ideology of "free markets" and "unfettered capitalism." This politics of ideology needs to be confronted upfront and the international trade regulations need to be seriously amended, so that the incentives for the global food industry to undertake massive land grabs in tropical countries are seriously undercut. Export of agricultural and mining goods from the tropical countries needs to be seriously taxed. Imposition of such environmental and carbon tariffs on the trade of goods produced through tropical deforestation could potentially reduce tropical deforestation, but such taxes and tariffs run against the usual free market ideology.

Confronting the politics of "free market" ideology, and the politics of knowledge in the form of de-marketization of climate governance, will also require tackling the politics of scale. Both free market ideology and unfettered capitalism assume that globalization is a win–win solution for all the players on this planet. This is truly a fantasy. The powerful industrialized countries, many of which have a long history of colonizing natural resources in Africa, Asia, and Latin America since the dawn of industrialization, might be winning through this game of globalization; however, the poor industrializing countries are the ultimate losers both in terms of the low-wage slave labour that they provide to serve the interests of global corporations and the huge environmental costs that are imposed on the southern countries through globalization practices. Promotion and protection of free market trading regime under the umbrella of WTO, at the core, is aimed at protecting the globalization agenda. This sheer politics of scale, whereby global scale corporations and industries can rob local scale communities of their natural and human resources, can be confronted through a massive rewriting of the WTO rules. All the countries of the south, and those industrializing countries that are serious about dealing with human induced global climate change, must immediately call for the rewriting of international trade rules to protect global to local environments, including tropical forests, oceans, lakes, rivers, and, above all, the global atmosphere that regulates the weather system on this planet.

Throughout the book, I have laid out various examples of politics of ideology, politics of knowledge, and the politics of scale in the climate governance regime. I have signalled the paths that will need to be followed from the global to local scale to confront these politics of ideology, knowledge and scale in the energy, transportation, and forestry sectors. Yet, the grandest challenge of all times for the human civilization resides in the "politics" of ideology, knowledge, and scale, whereby "powerful" industrial countries and transnational corporations are strongly resisting any meaningful policy and governance changes that potentially could solve the human induced global climate change problem. These powerful interest groups have maligned climate scientists through media campaigns such as "climate-gate." These interest groups are vehemently lobbying in the U.S. and EU capitals to stop any meaningful policy response to the climate change problem. A post-Kyoto climate governance regime, which confronts the politics of scale, ideology and knowledge, will provide equal voice to the relatively powerless vulnerable communities and industrializing countries. Strengthening the current UN system with a democratically elected global parliament could be a step in this direction that could potentially reduce global democracy deficit and enable the design of a more effective, legitimate, accountable and fairer climate policy regime. Yet, a global parliament could be hijacked by the powerful countries and transnational corporations, much like the U.S. Congress at a smaller scale. The co-optation of such a global parliament by powerful interest groups could be potentially even more dangerous than the current fragmented system of global governance by the UN agencies. However, there will be certain legitimacy and accountability built in such a global governance regime that is currently lacking at the global scale. A more legitimate and accountable global governance regime will also translate into a more legitimate and accountable post-Kyoto climate governance regime in the medium to long run.

References

Abbott, K.W. and D. Gartner. 2011. "The Green Climate Fund and the future of environmental governance." Earth System Governance working paper No. 16, Earth System Governance Project.

Achard, F., H.D. Eva, H.J. Stibig, P. Mayaux, J. Gallego, T. Richards, and J.P. Malingreau. 2002. "Determination of deforestation rates of the world's humid tropical forests." *Science* 297(5583): 999–1002.

Achard, F., H.D. Eva, P. Mayaux, H.J. Stibig, and A. Belward. 2004. "Improved estimates of net carbon emissions from land cover change in the tropics for the 1990s." *Global Biogeochemical Cycles* 18(2): 1–11.

Ackoff, R.L. 1974. *Redesigning the future: a systems approach to societal problems*: Wiley.

Adams, J.S. and T.O. McShane. 1992. *The myth of wild Africa: conservation without illusion*: Univ of California Pr.

Adaptation Fund. 2011a. "Adaptation Fund, Mr. Sven Harmeling." Adaptation Fund.

Adaptation Fund. 2011b. "Funded Projects." Adaptation Fund.

Adger, W.N. 2003. "Social capital, collective action, and adaptation to climate change." *Economic Geography* 79(4): 387–404.

Adger, W.N., T.A. Benjaminsen, K. Brown, and H. Svarstad. 2001. "Advancing a political ecology of global environmental discourses." *Development and Change* 32(4): 681–715.

Adger, W.N., N.W. Arnell, and E.L. Tompkins. 2005a. "Successful adaptation to climate change across scales." *Global Environmental Change Part A* 15(2): 77–86.

Adger, W.N., K. Brown, and E.L. Tompkins. 2005b. "The political economy of cross-scale networks in resource co-management." *Ecology and Society* 10(2): 9.

Adger, W.N., I. Lorenzoni, K.L. O'Brien, and WN Adgar. 2009. *Adapting to climate change: thresholds, values, governance*: Cambridge University Press.

Agarwal, A. 2001. "Making the Kyoto Protocol work." Center for Science and Environment.

Agnew, J. 1997. "The dramaturgy of horizons: geographical scale in the Reconstruction of Italy by the new Italian political parties, 1992–1995." *Political Geography* 16(2): 99–121.

Agranoff, R. and M. McGuire. 2001. "Big questions in public network management research." *Journal of Public Administration Research and Theory* 11(3): 295–326.

Agarwal, A. (n.d.) "Making the Kyoto protocol work." Online at www.cseindia.org/html/eyou/climate/pdf/cse_stat.pdf. Retrieved March 10, 2009.

Agrawal, A., and C.C. Gibson. 1999. "Enchantment and disenchantment: the role of community in natural resource conservation." *World Development* 27(4): 629–649.

Agrawal, A. and K. Redford. 2006. *Poverty, development, and biodiversity conservation: shooting in the dark?* Ann Arbor. 48109(734): 647–5948.

Albrecht, J., D. François, and K. Schoors. 2002. "A Shapley decomposition of carbon emissions without residuals." *Energy Policy* 30(9): 727–736.

Aldy, J.E. and R.N. Stavins. 2007. *Architectures for agreement: addressing global climate change in the post-Kyoto world*: Cambridge University Press.

Aldy, J.E. and R.N. Stavins. 2009. *Post-Kyoto international climate policy: implementing architectures for agreement: research from the Harvard Project on International Climate Agreements*: Cambridge University Press.

Aldy, J.E., S. Barrett, and R.N. Stavins. 2003. "Thirteen plus one: a comparison of global climate policy architectures." *Climate Policy* 3(4): 373–397.

Alzbutas, R. and A. Maioli. 2008. "Risk zoning in relation to risk of external events (application to IRIS design)." *International Journal of Risk Assessment and Management* 8(1): 104–122.

Amendola, A., Y. Ermoliev, T.Y. Ermolieva, V. Gitis, G. Koff, and J. Linnerooth-Bayer. 2000. "A systems approach to modeling catastrophic risk and insurability." *Natural Hazards* 21(2): 381–393.

Anderson, D. and R.H. Grove. 1990. *Conservation in Africa: peoples, policies and practice*: Cambridge University Press.

Antonakis, N. 1999. "Guns versus Butter." *Journal of Conflict Resolution* 43(4): 501–520.

Ariely, D. 2009. *Predictably irrational: the hidden forces that shape our decisions*: HarperCollins.

Arnell, N.W. 1984. "Flood hazard management in the United States and the National Flood Insurance Program." *Geoforum* 15(4): 525–542.

Arnell, N.W., M.J. Clark, and A.M. Gurnell. 1984. "Flood insurance and extreme events: the role of crisis in prompting changes in British institutional response to flood hazard." *Applied Geography* 4(2): 167–181.

Arsenault, C. 2010. "Seeing REDD on climate change: as the Cancun summit closes, some environmentalists say the REDD scheme is a boon for financers, not forests." Online at www.aljazeera.com/indepth/features/2010/12/201012919238402389.html.

Axelrod, R.M. 1976. *Structure of decision: the cognitive maps of political elites*: Princeton University Press.

Bäckstrand, K. 2008. "Accountability of networked climate governance: the rise of transnational climate partnerships." *Global Environmental Politics* 8(3): 74–102.

Baer, P. 2002. "Equity, greenhouse gas emissions, and global common resources." *Climate Change Policy: A Survey*: 393–408.

Bagstad, K.J., K. Stapleton, and J.R. D'Agostino. 2007. "Taxes, subsidies, and insurance as drivers of United States coastal development." *Ecological Economics* 63(2): 285–298.

Bailey, I. and G.A. Wilson. 2009. "Theorising transitional pathways in response to climate change: Technocentrism, ecocentrism, and the carbon economy." *Environment and Planning A* 41: 2324–2341.

Bakker, S., C. Haug, H. Van Asselt, J. Gupta, and R. Saidi. 2011. "The future of the CDM: same same, but differentiated?" *Climate Policy* 11(1): 752–767.

Baksi, S. and C. Green. 2007. "Calculating economy-wide energy intensity decline rate: the role of sectoral output and energy shares." *Energy Policy* 35(12): 6457–6466.

Banerjee, S. 2010. "Cancun open for green business, but REDD will destroy indigenous forest cultures." *Huffington Post*.

Bardach, E. and C. Lesser. 1996. "Accountability in human services collaboratives: for what? And to whom?" *Journal of Public Administration Research and Theory* 6(2): 197–224.

Barnard, J.R. 1978. "Externalities from urban growth: the case of increased storm runoff and flooding." *Land Economics* 54(3): 298–315.

Barnsley, I. 2008. *Reducing emissions from deforestation and forest degradation in developing countries (REDD): a guide for indigenous peoples.* UNU-IAS.

Barona, E., N. Ramankutty, G. Hyman, and O.T. Coomes. 2010. "The role of pasture and soybean in deforestation of the Brazilian Amazon." *Environmental Research Letters* 5: 024002.

Barrett, S. and R. Stavins. 2003. "Increasing participation and compliance in international climate change agreements." *International Environmental Agreements: Politics, Law and Economics* 3(4): 349–376.

Bateson, G. 1980. *Mind and nature: a necessary unity*: Bantam Books.

Batty, M. 2007. *Cities and complexity: understanding cities with cellular automata, agent-based models, and fractals*: MIT Press.

Bazzani, G.M. 2005. "A decision support for an integrated multi-scale analysis of irrigation: DSIRR." *Journal of Environmental Management* 77(4): 301–314.

Becker, G.S. 1976. *The economic approach to human behavior*: University of Chicago Press.

Becker, G.S. 1993. "Nobel lecture: the economic way of looking at behavior." *Journal of Political Economy*: 385–409.

Beckerman, W. 1994. "'Sustainable development': is it a useful concept?" *Environmental Values* 3(3): 191–209.

Behn, R.D. 2001. *Rethinking democratic accountability*: Brookings Institution Press.

Bell, H.M. and G.A. Tobin. 2007. "Efficient and effective? The 100-year flood in the communication and perception of flood risk." *Environmental Hazards* 7(4): 302–311.

Bell, L. 2011. *Climate of corruption: politics and power behind the global warming hoax*: Greenleaf Book Group Press.

Benedick, R.E. 2001. "Striking a new deal on climate change." *Issues in Science and Technology* 18(1): 71–76.

Benner, T., W.H. Reinicke, and J.M. Witte. 2004. "Multisectoral networks in global governance: towards a pluralistic system of accountability." *Government and Opposition* 39(2): 191–210.

Berkes, F. 2002. "Cross-scale institutional linkages: perspectives from the bottom up." *The Drama of the Commons*: 293–321.

Berkes, F. 2006. "From community-based resource management to complex systems: the scale issue and marine commons." *Ecology and Society* 11(1): 45.

Biermann, F. 2005. "Between the USA and the south: strategic choices for European climate policy." *Climate Policy* 5(3): 273–290.

Biermann, F., P. Pattberg, H. Van Asselt, and F. Zelli. 2009. "The fragmentation of global governance architectures: a framework for analysis." *Global Environmental Politics* 9(4): 14–40.

Biermann, F., P. Pattberg, and F. Zelli. 2010. *Global climate governance beyond 2012: architecture, agency and adaptation*: Cambridge University Press.

Biermann, F., K. Abbott, S. Andresen, K. Bäckstrand, S. Bernstein, M.M. Betsill, H. Bulkeley, B. Cashore, J. Clapp, C. Folke, A. Gupta, J. Gupta, P.M. Haas, A. Jordan, N. Kanie, T. Kluvánková-Oravská, L. Lebel, D. Liverman, J. Meadowcroft, R.B. Mitchell, P. Newell, S. Oberthür, L. Olsson, P. Pattberg, R. Sánchez-Rodríguez, H. Schroeder,

A. Underdal, C. Vogel, S. Camargo Vieira, O.R. Young, A. Brock, and R. Zondervan. 2012. "Navigating the Anthropocene: improving earth system governance." *Science* 335:1306–1307.

Bin, O., J.B. Kruse, and C.E. Landry. 2008. "Flood hazards, insurance rates, and amenities: evidence from the coastal housing market." *Journal of Risk and Insurance* 75(1): 63–82.

Bingham, S. 2007. 'Climate change: a moral issue.' In S. Moser and L. Dilling (eds.), *Creating a climate for change: communicating climate change and facilitating social change*, 153–66: Cambridge University Press.

Blaikie, P. and S. Jeanrenaud. 1997. "Biodiversity and human welfare." *Social Change and Conservation*: 46–70.

Blanchard-Boehm, R.D., K.A. Berry, and P.S. Showalter. 2001. "Should flood insurance be mandatory? Insights in the wake of the 1997 New Year's Day flood in Reno-Sparks, Nevada." *Applied Geography* 21(3): 199–221.

Blom, B., T. Sunderland, and D. Murdiyarso. 2010. "Getting REDD to work locally: lessons learned from integrated conservation and development projects." *Environmental Science and Policy* 13(2): 164–172.

Bodansky, D., S. Chou, and C. Jorge-Tresolini. 2004. "International climate efforts beyond 2012: a survey of approaches." Pew Center on Global Climate Change.

Bode, S. 2004. "Equal emissions per capita over time: a proposal to combine responsibility and equity of rights for post-2012 GHG emission entitlement allocation." *European Environment* 14(5): 300–316.

Bogason, P. and J.A. Musso. 2006. "The democratic prospects of network governance." *The American Review of Public Administration* 36(1): 3–18.

Bord, R.J., R.E. O'Connor, and A. Fisher. 2000. "In what sense does the public need to understand global climate change?" *Public Understanding of Science* 9(3): 205–218.

Boruch, R.F. and A. Petrosino. 2004. "Meta-analysis, systematic reviews, and research syntheses." *Handbook of Practical Program Evaluation* 19: 176.

Bostrom, A., M.G. Morgan, B. Fischhoff, and D. Read. 1994. "What do people know about global climate change? 1. Mental models." *Risk Analysis* 14(6): 959–970.

Botzen, W.J.W., J. Aerts, and J. Van Den Bergh. 2009. "Willingness of homeowners to mitigate climate risk through insurance." *Ecological Economics* 68(8–9): 2265–2277.

Boucher, D. 2008. *Out of the woods: a realistic role for tropical forests in curbing global warming*: Union of Concerned Scientists.

Boyle, G., J. Kirton, R.M. Lof, and T. Nayler. 2009. "Transitioning from the CDM to a Clean Development Fund." *Climate Law Review* 3(1): 16–24.

Branson, R., P. Cosier, G. Dyer, T. Flannery, B. Foran, D. Foster, I. Lowe, A. Mitchell, B. Pittock, and G. Russell. 2008. "Now or never." *Quarterly Essay* (32): 92 ff.

Brenner, N. 2001. "The limits to scale? Methodological reflections on scalar structuration." *Progress in Human Geography* 25(4): 591–614.

Brockington, D. and K. Schmidt-Soltau. 2004. "The social and environmental impacts of wilderness and development." *Oryx* 38(2): 140–142.

Brockington, D., J. Igoe, and K. Schmidt-Soltau. 2006. "Conservation, human rights, and poverty reduction." *Conservation Biology* 20(1): 250–252.

Bromley, D.W. 1998. "Searching for sustainability: the poverty of spontaneous order." *Ecological Economics* 24(2): 231–240.

Bromley, D.W. and A. Vatn. 1994. "Choices without prices without apologies." *Journal of Environmental Economics and Management* 26: 129–148.

Brosius, J.P. 2006. "Between politics and poetics: narratives of dispossession in Sarawak, East Malaysia." *Reimagining Political Ecology*. Duke University Press: 281–322.

Brown, C.J., and M. Purcell. 2005. "There's nothing inherent about scale: political ecology, the local trap, and the politics of development in the Brazilian Amazon." *Geoforum* 36(5): 607–624.

Brown, K. 2002. "Innovations for conservation and development." *The Geographical Journal* 168(1): 6–17.

Brunner, R.D. and A.H. Lynch. 2010. *Adaptive governance and climate change*: American Meteorological Society.

Bruntland, G. 1987. *Our common future: the world commission on environment and development*: Oxford University Press.

Bryan, E., W. Akpalu, M. Yesuf, and C. Ringler. 2010. "Global carbon markets: opportunities for sub-Saharan Africa in agriculture and forestry." *Climate and Development* 2(4): 309–331.

Burby, R.J. 2001. "Flood insurance and floodplain management: the US experience." *Global Environmental Change Part B: Environmental Hazards* 3(3–4): 111–122.

Burby, R.J. 2006. "Hurricane Katrina and the paradoxes of government disaster policy: bringing about wise governmental decisions for hazardous areas." *Annals of the American Academy of Political and Social Science* 604(1): 171–191.

Burnett, H.S. 2009. "Carbon offsets scam." *Washington Times*, March 8, 2009.

Butler, R.A., L.P. Koh, and J. Ghazoul. 2009. "REDD in the red: palm oil could undermine carbon payment schemes." *Conservation Letters* 2: 67–73.

Cadman, T. and T.N. Maraseni. 2011. "The governance of climate change: evaluating the governance quality and legitimacy of the United Nations' REDD-plus programme." *The International Journal of Climate Change: Impacts and Responses* 2(3): 103–124.

Caputo, D.A. 1975. "New perspectives on the public policy implications of defense and welfare expenditures in four modern democracies: 1950–1970." *Policy Sciences* 6(4): 423–446.

Carolan, M.S. 2007. "One step forward, two steps back: flood management policy in the United States." *Environmental Politics* 16(1): 36–51.

Carrubba, C.J. and A. Singh. 2004. "A decision theoretic model of public opinion: guns, butter, and European common defense." *American Journal of Political Science* 48(2): 218–231.

Carvalho, A. 2007. "Ideological cultures and media discourses on scientific knowledge: re-reading news on climate change." *Public Understanding of Science* 16(2): 223–243.

Cash, D.W., W.N. Adger, F. Berkes, P. Garden, L. Lebel, P. Olsson, L. Pritchard, and O. Young. 2006. "Scale and cross-scale dynamics: governance and information in a multi-level world." *Ecology and Society* 11(2): 8.

Cass, L.R. 2006. *The failures of American and European climate policy: international norms, domestic politics, and unachievable commitments*: State University of New York Press.

Chakravarty, S., A. Chikkatur, H. De Coninck, S. Pacala, R. Socolow, and M. Tavoni. 2009. "Sharing global CO_2 emission reductions among one billion high emitters." *Proceedings of the National Academy of Sciences* 106(29): 11884–11888.

Chandani, A., Harmeling, S., and Kaloga, A.O. (eds.) 2009. *The Adaptation Fund: a model for the future?* International Institute for Environment and Development.

Chao, P.T., J.L. Floyd, and W. Holliday. 1998. "Empirical studies of the effect of flood risk on housing prices." DTIC Document.

Chen, K., R. Blong, and C. Jacobson. 2003. "Towards an integrated approach to natural hazards risk assessment using GIS: with reference to bushfires." *Environmental Management* 31(4): 546–560.

Chhatre, A. and A. Agrawal. 2009. "Trade-offs and synergies between carbon storage and livelihood benefits from forest commons." *Proceedings of the National Academy of Sciences* 106(42): 17667–17670.

Chivers, J. and N.E. Flores. 2002. "Market failure in information: the national flood insurance program." *Land Economics* 78(4): 515–521.

Christensen, G.L. and J.C. Olson. 2002. "Mapping consumers' mental models with ZMET." *Psychology and Marketing* 19(6): 477–501.

Chung, E.S. and K.S. Lee. 2009. "Prioritization of water management for sustainability using hydrologic simulation model and multicriteria decision making techniques." *Journal of Environmental Management* 90(3): 1502–1511.

Churchman, C.W. 1967. "Guest Editorial: Wicked Problems." *JSTOR*.

Ciplet, D., J.T. Roberts, M. Khan, L. He, S. Fields, S. Huq, J. Ayers, S. Anderson, A. Chandani, and H. Reid. 2011. "Adaptation finance: how can Durban deliver on past promises?" IIED Briefing Paper. Online at http://pubs.iied.org/17115IIED.html.

Clark, D.H. 2001. "Trading butter for guns." *Journal of Conflict Resolution* 45(5): 636–660.

Clark, B. and J.B. Foster. 2009. "Ecological imperialism and the global metabolic rift: unequal exchange and the Guano/Nitrates trade." *International Journal of Comparative Sociology* 50: 311–334.

Clayton, J.L. 1976. "The fiscal limits of the warfare-welfare state: defense and welfare spending in the United States since 1900." *The Western Political Quarterly*: 364–383.

Cleaver, F. 1999. "Paradoxes of participation: questioning participatory approaches to development." *Journal of International Development* 11(4): 597–612.

Climate Action Network. 2007. "Reducing emissions from deforestation and forest degradation (REDD): A discussion paper for the Climate Action Network (CAN)." Climate Action Network.

Coelho, J. 2011. "Bribery, collusion hinder U.N. carbon scheme: research." Reuters.

Coen, D. and M. Thatcher. 2008. "Network governance and multi-level delegation: European networks of regulatory agencies." *Journal of Public Policy* 28(01): 49–71.

Comfort, L., B. Wisner, S. Cutter, R. Pulwarty, K. Hewitt, A. Oliver-Smith, J. Wiener, M. Fordham, W. Peacock, and F. Krimgold. 1999. "Reframing disaster policy: the global evolution of vulnerable communities." *Environmental Hazards* 1(1): 39–44.

Conant, J. 2011. "Do trees grow on money?" *Earth Island Journal*, Autumn.

CoolNRG. 2009. "What is the Clean Development Mechanism?" CoolNRG.

Cooney, D. 2011. "3 sticking points to tackle on REDD+ in Durban, says facilitator." *Forests news: a blog*, Center for International Forestry Research.

Cooper, R. 2001. "The Kyoto Protocol: a flawed concept." Working paper. Online at www.feem.it/userfiles/attach/Publication/NDL2001/NDL2001-052.pdf.

Cooper, R.N. 1998. "Toward a real global warming treaty." *Foreign Affairs*: 66–79.

Corbera, E. and H. Schroeder. 2011. "Governing and implementing REDD+." *Environmental Science and Policy* 14(2): 89–99.

Corbera, E., M. Estrada, P. May, G. Navarro, and P. Pacheco. 2011. "Rights to land, forests and carbon in REDD+: insights from Mexico, Brazil and Costa Rica." *Forests* 2(1): 301–342.

Cotula, L., S. Vermeulen, R. Leonard, and J. Keeley. 2009. "Land grab or development opportunity? Agricultural investment and international land deals in Africa." IIED/FAO/IFAD, London/Rome.

Cox, G. 2010. "The clean development mechanism as a vehicle for technology transfer and sustainable development: myth or reality?" *Law, Environment and Development Journal* 6(2): 179–199.

Cummins, J.D. 2006. "Should the government provide insurance for catastrophes?" *Federal Reserve Bank of St. Louis Review* 88(4): 337–379.

Daly, H.E. and J. Farley. 2010. *Ecological economics: principles and applications*: Island Press.

Damianos, D., L.A. Shabman, Virginia Polytechnic Institute, and State University. Water Resources Research Center. 1976. *Land prices in flood hazard areas: applying methods of land value analysis*: Virginia Water Resources Research Center, Virginia Polytechnic Institute and State University.

Daniel, H. 2001. "Replenishment versus retreat: the cost of maintaining Delaware's beaches." *Ocean and Coastal Management* 44(1–2): 87–104.

Daniel, S. and A. Mittal. 2010. *(Mis)investment in agriculture: the role of the international finance corporation in global land grabs*: The Oakland Institute.

Danum, Y.P., Jasoil, Y.M. Putih, and Y.R. Bambu. 2011. "Rights, Forests and Climate Change: REDD+ in Indonesia." Forest Peoples Programme. Online at www.forestpeoples.org/topics/redd-and-related-initiatives/publication/2011/fpp-series-rights-forests-and-climate-october-2.

De Coninck, H., C. Fischer, R.G. Newell, and T. Ueno. 2008. "International technology-oriented agreements to address climate change." *Energy Policy* 36(1): 335–356.

De Soto Blass, Maria Luisa Fernandez. 2010 "The fiscal implications of the Clean Development Mechanism." *The Business Review* 15(2): 253–258.

DeFries, R.S., R.A. Houghton, M.C. Hansen, C.B. Field, D. Skole, and J. Townshend. 2002. "Carbon emissions from tropical deforestation and regrowth based on satellite observations for the 1980s and 1990s." *Proceedings of the National Academy of Sciences* 99(22): 14256.

DeFries, S., T. Rudel, M. Uriarte, and M. Hansen. 2010. "Deforestation driven by urban population growth and agricultural trade in the twenty-first century." *Nature Geoscience*: DOI: 10.1038/NGEO1756.

Dehring, C. 2006a. "The value of building codes." *Regulation* 29(2): 10–13.

Dehring, C.A. 2006b. "Building codes and land values in high hazard areas." *Land Economics* 82(4): 513–528.

Delaney, D. and H. Leitner. 1997. "The political construction of scale." *Political Geography* 16(2): 93–98.

Den Elzen, M., J. Fuglestvedt, N. Höhne, C. Trudinger, J. Lowe, B. Matthews, B. Romstad, C.P. de Campos, and N. Andronova. 2005. "Analysing countries' contribution to climate change: scientific and policy-related choices." *Environmental Science and Policy* 8(6): 614–636.

Dessler, A.E. and E. Parson. 2010. *The science and politics of global climate change: a guide to the debate*: Cambridge University Press.

Dewey, John. 1927. *The public and its problems*: Henry Holt and Company.

Dinar, A., S.M. Rahman, D.F. Larson, and P. Ambrosi. 2011. "Local actions, global impacts: international cooperation and the CDM." *Global Environmental Politics* 11(4): 108–133.

Donnelly, W.A. 1989. "Hedonic price analysis of the effect of a floodplain on property values." *Journal of the American Water Resources Association* 25(3): 581–586.

Dowdle, M.W. 2006. *Public accountability: designs, dilemmas and experiences*: Cambridge University Press.

Duan, M. 2011. "Reform of the Clean Development Mechanism: where should we head for?" *Carbon and Climate Law Review* 5(2): 169–177.

Duic, N., L.M. Alves, F. Chen, and M. da Graça Carvalho. 2003. "Potential of Kyoto Protocol Clean Development Mechanism in transfer of clean energy technologies to small island developing states: case study of Cape Verde." *Renewable and Sustainable Energy Reviews* 7(1): 83–98.

Durant, J.R., G.A. Evans, and G.P. Thomas. 1989. "The public understanding of science." *Nature* 340: 11–14.

Durfee, J. 2006. "'Social Change' and 'Status Quo' Framing Effects on Risk Perception." *Science Communication* 27(4): 459.

Duro, J.A. and E. Padilla. 2006. "International inequalities in per capita CO_2 emissions: a decomposition methodology by Kaya factors." *Energy Economics* 28(2): 170–187.

Eagleton, T. 1991. *Ideology: an introduction*: Verso Books.

Ehrlich, P.R., and J.P. Holdren. 1971. "Impact of population growth." *Science* 171(3977): 1212–1217.

Ekins, P. 2004. "Step changes for decarbonising the energy system: research needs for renewables, energy efficiency and nuclear power." *Energy Policy* 32(17): 1891–1904.

Escobar, A. 2001. "Culture sits in places: reflections on globalism and subaltern strategies of localization." *Political Geography* 20(2): 139–174.

Evans, G. and J. Durant. 1995. "The relationship between knowledge and attitudes in the public understanding of science in Britain." *Public Understanding of Science* 4(1): 57–74.

Evans, M., S. Legro, and I. Popov. 2000. "The climate for joint implementation: case studies from Russia, Ukraine, and Poland." *Mitigation and Adaptation Strategies for Global Change* 5(4): 319–336.

Evatt, D.S. 2000. "Does the national flood insurance program drive floodplain development? A review of the evidence." *Journal of Insurance Regulation* 18(4): 497–526.

Fagerlin, A., C. Wang and P.A. Ubel. 2005. "Reducing the influence of anecdotal reasoning on people's health care decisions: is a picture worth a thousand statistics?" *Medical Decision Making* 25(4): 398–405.

Faith, D.P. and P.A. Walker. 1996. "Integrating conservation and development: effective trade-offs between biodiversity and cost in the selection of protected areas." *Biodiversity and Conservation* 5(4): 431–446.

Fearnside, P.M. 2000. "Global warming and tropical land-use change: greenhouse gas emissions from biomass burning, decomposition and soils in forest conversion, shifting cultivation and secondary vegetation." *Climatic Change* 46(1): 115–158.

Fell, R., J. Corominas, C. Bonnard, L. Cascini, E. Leroi, and W.Z. Savage. 2008. "Guidelines for landslide susceptibility, hazard and risk zoning for land use planning." *Engineering Geology* 102(3–4): 85–98.

Flåm, K.H. and J.B. Skjaerseth. 2009. "Does adequate financing exist for adaptation in developing countries?" *Climate Policy* 9(1): 109–114.

Flannery, T. 2010. *Now or never: why we need to act now to achieve a sustainable future*: HarperCollins.

Forests Dialogue. 2008. *Beyond REDD: the role of forests in climate change*: Forests Dialogue.

Fortum Corporation, Corporate Communications Media Room 2011. "Russia Approved Two Fortum JOint Implementation Projects." Fortum Corporation.

Fotheringham, A.S. and D.W.S. Wong. 1991. "The modifiable areal unit problem in multivariate statistical analysis." *Environment and Planning A* 23(7): 1025–1044.

Fox, C.R. and A. Tversky. 1995. "Ambiguity aversion and comparative ignorance." *Quarterly Journal of Economics* 110: 585–603.

Franks, P. and T. Blomley. 2004. "Fitting ICD into a project framework: a CARE perspective." *Getting Biodiversity Projects to Work: Towards More Effective Conservation and Development*: 77–97.

Frederick, S., G. Loewenstein, and T. O'Donoghue. 2002. "Time discounting and time preference: a critical review." *Journal of Economic Literature* 40(2): 351–401.

Frederickson, D.G. and H.G. Frederickson. 2006. *Measuring the performance of the hollow state*: Georgetown University Press.

Freeman, A.M. 2003. *The measurement of environmental and resource values: theory and methods*: RFF press.

Friedman, R.M., S.V. Dunn, and W.J. Merrell, Jr. 2002. "Summary of the Heinz Center report on coastal erosion and the national flood insurance program." *Journal of Coastal Research*: 568–575.

Frumkin, H., J. Hess, G. Luber, J. Malilay, and M. McGeehin. 2008. "Climate change: the public health response." *American Journal of Public Health* 98(3): 435.

Ghazoul, J., R.A. Butler, J. Mateo-Vega, and L.P. Koh. 2010. "REDD: a reckoning of environment and development implications." *Trends in Ecology and Evolution* 25(7): 396–402.

Giddens, A. 2009. *The politics of climate change*: Wiley Online Library.

Gigerenzer, G., W. Hell, and H. Blank. 1988. "Presentation and content: the use of base rates as a continuous variable." *Journal of Experimental Psychology: Human Perception and Performance* 14(3): 513–525.

Gigerenzer, G., R. Hertwig, E. van den Broek, B. Fasolo, and K.V. Katsikopoulos. 2005. "'A 30% chance of rain tomorrow': How does the public understand probabilistic weather forecasts?" *Risk Analysis* 25: 623–629.

Gilovich, T., D.W. Griffin, and D. Kahneman. 2002. *Heuristics and biases: the psychology of intuitive judgement*: Cambridge University Press.

Glantz, M. 1977. "Nine fallacies of natural disaster: the case of the Sahel." *Climatic Change* 1(1): 69–84.

Glantz, M.H. 2003. *Climate affairs: a primer*: Island Pr.

Global Environmental Subcommittee of the Environmental Committee of the Industrial Structure Council. 2003. "Interim Report: Perspectives and Actions to Construct a Future Sustainable Framework on Climate Change." Industrial Structure Council.

Golding, D., S. Krimsky, and A. Plough. 1992. "Evaluating risk communication: narrative vs. technical presentations of information about radon." *Risk Analysis* 12.1: 27–35.

Goldsmith, S. and W.D. Eggers. 2004. *Governing by network: the new shape of the public sector*: Brookings Institution Press.

Gopalakrishnan, C. and N. Okada. 2007. "Designing new institutions for implementing integrated disaster risk management: key elements and future directions." *Disasters* 31(4): 353–372.

Gosling, Melanie. 2011. "Climate: there's still time to make a change." IOL South Africa.

Government Accountability Office. 2001. "Flood insurance information on the financial condition of the National Flood Insurance Program, statement of Stanley J. Czerwinski, Director, Physical Infrastructure Issues." Government Accountability Office.

Grasso, M. 2011. "The role of justice in the north–south conflict in climate change: the case of negotiations on the Adaptation Fund." *International Environmental Agreements: Politics, Law and Economics* 11(4): 361–377.

Green, R. and M. Petal. 2008. "Stocktaking report and policy recommendations on risk awareness and education on natural catastrophes." *OECD Journal: General Papers* 2008(3): 217–305.

Gregory, R. and R.L. Keeney. 1994. "Creating policy alternatives using stakeholder values." *Management Science*: 1035–1048.

Griffith, C.T. 1994. "The National Flood Insurance Program: unattained purposes, liability in contract, and takings." *Wm and Mary L. Rev.* 35: 727–1801.

Griffiths, T. 2009. "Seeing 'RED'? Forests, climate change mitigation and the rights of indigenous peoples and local communities." Forest People's Programme. Online at www.rightsandresources.org/publication_details.php?publicationID=923.

Gruber, J.E. 1987. *Controlling bureaucracies: dilemmas in democratic governance*: University of California Press.

Guesnerie, R. 2006. "The design of post-Kyoto climate schemes: an introductory analytical assessment." Working Paper 2006–11.

Guikema, S. and M. Milke. 1999. "Quantitative decision tools for conservation programme planning: practice, theory and potential." *Environmental Conservation* 26(3): 179–189.

Gupta, J. 2003. "Engaging developing countries in climate change: KISS and make-up!" In David Michel (ed.), *Climate policy for the 21st century: meeting the long-term challenge of global warming*: Center for Transatlantic Relations.

Haas, P.M. 2004. "Addressing the global governance deficit." *Global Environmental Politics* 4(4): 1–15.

Habermas, J. 1998. *Between facts and norms: contributions to a discourse theory of law and democracy*: The MIT Press.

Habermas, J. and T. McCarthy. 1985. *The theory of communicative action: reason and the rationalization of society*: Beacon Pr.

Haites, E. and F. Yamin. 2000. "The Clean Development Mechanism: proposals for its operation and governance." *Global Environmental Change* 10(1): 27–45.

Hämäläinen, R.P. and S. Alaja. 2008. "The threat of weighting biases in environmental decision analysis." *Ecological Economics* 68(1): 556–569.

Hameed, K. (ed.) 2005. "Flood management in Pakistan: a report prepared for UNDP." UNDP.

Hansen, R. and D. Bausch. 2005. "A GIS-based methodology for exporting the hazards U.S. (hazus) earthquake model for global applications." Federal Emergency Management Agency. Online at www.usehazus.com/docs/gis_global_hazus_paper.pdf, 64 pp. Retrieved February 27, 2012.

Hansen, R., and D. Bausch. 2006. "A GIS-based methodology for exporting the Hazards US (HAZUS) earthquake model for global applications."

Hansen, J., Mki. Sato, and R. Ruedy, 2012. "Perception of climate change." *Proc. Natl. Acad. Sci.* 109: 14726–14727, E2415-E2423, doi:10.1073/pnas.1205276109.

Harinck, F., E. Van Dijk, I. Van Beest, and P. Mersmann. 2007. "When gains loom larger than losses: reversed loss aversion for small amounts of money." *Psychological Science* 18(12): 1099–1105.

Harrison, D.M., G.T. Smersh, and A.L. Schwartz. 2001. "Environmental determinants of housing prices: the impact of flood zone status." *Journal of Real Estate Research* 21(1): 3–20.

Harrison, K., and L.M.I. Sundstrom. 2007. "The comparative politics of climate change." *Global Environmental Politics* 7(4): 1–18.

Hartman, S.W. 1973. "The impact of defense expenditures on the domestic American economy 1946–1972." *Public Administration Review* 33(4): 379–390.

Haynes, P. 2003. *Managing complexity in the public services*: Open University Press.

Heinrich, C.J. and L.E. Lynn. 2000. *Governance and performance: new perspectives*: Georgetown University Press.

Heintz, R.J. and R.S.J. Tol. 1995. "Joint implementation and uniform mixing." *Energy Policy* 23(10): 911–917.

Helm, D. and C. Hepburn. 2009. *The economics and politics of climate change*: Oxford University Press.

Helvarg, D. 2005. "FEMA's next disaster." *Multinational Monitor* 26(9/10): 16–17.

Herath, G. 2004. "Incorporating community objectives in improved wetland management: the use of the analytic hierarchy process." *Journal of Environmental Management* 70(3): 263–273.

Hirsch, P.D., W.M. Adams, J.P. Brosius, A. Zia, N. Bariola, and J.L. Dammert. 2011. "Acknowledging conservation trade-offs and embracing complexity." *Conservation Biology* 25(2): 259–264.

Hisschemöller, M. and R. Hoppe. 1995. "Coping with intractable controversies: the case for problem structuring in policy design and analysis." *Knowledge, Technology, and Policy* 8(4): 40–60.

Höhne, N., M. den Elzen and M. Weiss. 2006. "Common but differentiated convergence (CDC): a new conceptual approach to long-term climate policy." *Climate Policy* 6: 181–199.

Holway, J.M. and R.J. Burby. 1990. "The effects of floodplain development controls on residential land values." *Land Economics* 66(3): 259–271.

Holway, J.M. and R.J. Burby. 1993. "Reducing flood losses local planning and land use controls." *Journal of the American Planning Association* 59(2): 205–216.

Holzknecht, K. 2011. "How to connect REDD+ and markets while avoiding crises over access to land" *Forest news: a blog* by the Center for International Forestry Research, Bogor, Indonesia.

Horstmann, B. 2011. "Operationalizing the Adaptation Fund: challenges in allocating funds to the vulnerable." *Climate Policy* 11(4): 1086–1096.

Houghton, RA. 2003a. "Why are estimates of the terrestrial carbon balance so different?" *Global Change Biology* 9(4): 500–509.

Houghton, R.A. 2003b. "Revised estimates of the annual net flux of carbon to the atmosphere from changes in land use and land management 1850–2000." *Tellus B* 55(2): 378–390.

Howarth, D. 2009. "Power, discourse, and policy: articulating a hegemony approach to critical policy studies." *Critical Policy Studies* 3(3–4): 309–335.

Howarth, R.B. and M.A. Wilson. 2006. "A theoretical approach to deliberative valuation: aggregation by mutual consent." *Land Economics* 82(1): 1–16.

Howitt, R. 1998. "Scale as relation: musical metaphors of geographical scale." *Area* 30(1): 49–58.

Hulme, D. and M. Murphree. 1999. "Communities, wildlife and the 'new conservation' in Africa." *Journal of International Development* 11(2): 277–285.

Hulme, D. and M. Murphree. 2001. *African wildlife and livelihoods: the promise and performance of community conservation*: James Currey Ltd.

Hulme, M. 2009. *Why we disagree about climate change: understanding controversy, inaction and opportunity*: Cambridge University Press.

Huntington, H.P. 2007. "Creating a climate for change: communicating climate change and facilitating social change." *Ecoscience* 14(4): 545–546.

Hurtado, M.E. 2011. "Eight countries get Climate Investment Funds." Science and Development Network, Santiago.

International Energy Agency. 2006. *CO_2 emissions from fuel combustion: 1971–2004*: OECD/IEA.

IPCC (Intergovernmental Panel on Climate Change). 1995. *Climate change 1995: IPCC second assessment.* Online at http://www.ipcc.ch/pdf/climate-changes-1995/ipcc-2nd-assessment/2nd-assessment-en.pdf

IPCC. 2001. *Climate change 2001: synthesis report, contribution of Working Groups I, II and III to the third assessment report of the Intergovernmental Panel on Climate Change*: IPCC. [Core writing team, David J. Dokken, Maria Noguer, Paul van der Linden, Cathy Johnson, and Jiahua Pan (eds.).]

IPCC. 2007. *Climate change 2007: synthesis report, contribution of Working Groups I, II and III to the fourth assessment report of the Intergovernmental Panel on Climate Change*: IPCC. [Core writing team: R.K. Pachauri, and A. Reisinger (eds.).]

IPCC. 2012 *Managing the risks of extreme events and disasters to advance climate change adaptation*: Cambridge University Press. [C.B. Field, V. Barros, T.F. Stocker, D. Qin, D.J. Dokken, K.L. Ebi, M.D. Mastrandrea, K.J. Mach, S.K. Allen, G.-K. Plattner, M. Tignor, and P.M. Midgley (eds.).]

Isenberg, J. and C. Potvin. 2010. "Financing REDD in developing countries: a supply and demand analysis." *Climate Policy* 10(2): 216–231.

Jackson, J. 2011. "UN Clean Development Mechanism Hits 3,000 Projects." *Earth Times*, May 8, 2011.

Jacobi, S.K. and B.F. Hobbs. 2007. "Quantifying and mitigating the splitting bias and other value tree-induced weighting biases." *Decision Analysis* 4(4): 194–210.

Jacoby, H.D., R.G. Prinn, and R. Schmalensee. 1998. "Kyoto's unfinished business." *Foreign Aff.* 77: 54.

Jacoby, H.D., R. Schmalensee, and I. Sue Wing. 1999. "Toward a useful architecture for climate change negotiations." Report No. 49. MIT Joint Program on the Science and Policy of Global Change. Online at http://dspace.mit.edu/bitstream/handle/1721.1/3598/MITJPSPGC_Rpt49.pdf?sequence=1.

Jenni, K.E. and G. Loewenstein. 1997. "Explaining the 'identifiable victim effect'." *Journal of Risk and Uncertainty* 14: 235–57.

Jonas, A.E.G. 1994. "The scale politics of spatiality." *Environment and Planning D* 12: 257–264.

Jones, C., W.S. Hesterly, and S.P. Borgatti. 1997. "A general theory of network governance: exchange conditions and social mechanisms." *Academy of Management Review*: 911–945.

Kahneman, D. and A. Tversky. 1973. "On the psychology of prediction." *Psychological Review 80*(4): 237–251.

Kameyama, Y. 2004. "The future climate regime: a regional comparison of proposals." *International Environmental Agreements: Politics, Law, and Economics* 4(4): 307–326.

Kanninen, M., D. Murdiyarso, F. Seymour, A. Angelsen, S. Wunder, and L. German. 2007. *Trees grow on money? The implications of deforestation research for policies to promote REDD*: Center for International Forestry Research.

Karsenty, A. and S. Ongolo. 2011. "Can 'fragile states' decide to reduce their deforestation? The inappropriate use of the theory of incentives with respect to the REDD mechanism." Forest Policy and Economics.

Kauffman, S.A. 1993. *The origins of order: self-organization and selection in evolution*: Oxford University Press, USA.

Kauffman, S.A. 1995. *At home in the universe: the search for laws of self-organization and complexity*: Oxford University Press, USA.

Kawase, R., Y. Matsuoka and J. Fujino. 2006. "Decomposition analysis of CO_2 emission in long-term climate stabilization scenarios." *Energy Policy* 34(15): 2113–2122.

Kaya, Y. 1990. "Impact of carbon dioxide emission control on GNP growth: interpretation of proposed scenarios." IPCC Energy and Industry Subgroup, Response Strategies Working Group, Paris, 76.

Keeler, A., W. Kriesel, and C. Landry. 2003. "Expanding the National Flood Insurance Program to cover coastal erosion damage." *Journal of Agricultural and Applied Economics* 35(3): 639–648.

Keeney, R.L. 1988. "Building models of values." *European Journal of Operational Research* 37(2): 149–157.

Keeney, R.L. 1996a. *Value-focused thinking: a path to creative decisionmaking*: Harvard University Press.

Keeney, R.L. 1996b. "Value-focused thinking: identifying decision opportunities and creating alternatives." *European Journal of Operational Research* 92(3): 537–549.

Keeney, R.L. 2002. "Common mistakes in making value trade-offs." *Operations Research*: 935–945.

Keeney, R.L. and T.L. McDaniels. 1999. "Identifying and structuring values to guide integrated resource planning at BC Gas." *Operations Research*: 651–662.

Kelman, S. 1981. "Cost-benefit analysis." *Regulation*: 33.

Kelman, S. 2006. "Improving service delivery performance in the United Kingdom: organization theory perspectives on central intervention strategies." *Journal of Comparative Policy Analysis* 8(4): 393–419.

Kempton, W. 1991. "Lay perspectives on global climate change." *Global Environmental Change* 1(3): 183–208.

Kempton, W. 1997. "How the public views climate change." *Environment: Science and Policy for Sustainable Development* 39(9): 12–21.

Kendall, G. (2008). "Civic exchange" [television news broadcast]. In *The Daily Motion*. Hong Kong: FORAtv. Retrieved from www.dailymotion.com/video/xgk4cj_gail-

Kerr, J., C. Foley, K. Chung, and R. Jindal. 2006. "Reconciling environment and development in the Clean Development Mechanism." *Journal of Sustainable Forestry* 23(1): 1–18.

Kahneman, D. and A. Tversky. 1979. "Prospect theory: an analysis of decision under risk." *Econometrica* 47: 263–291.

Kauffman, S.A. 1995. *At home in the universe: the search for the laws of the self organization and complexity*: Oxford University Press.

Kickert, W.J.M., E.H. Klijn, and J.F.M. Koppenjan. 1997. *Managing complex networks: strategies for the public sector*: Sage Publications Ltd.

Kiker, G.A., T.S. Bridges, A. Varghese, T.P. Seager, and I. Linkov. 2005. "Application of multicriteria decision analysis in environmental decision making." *Integrated Environmental Assessment and Management* 1(2): 95–108.

Kim, Y.G. and K.A. Baumert. 2002. Y. Kim and K. Baumert 2002 "Reducing uncertainty through dual intensity targets", in K. Baumert (ed.), *Building on the Kyoto Protocol: options for protecting the climate*, World Resources Institute, Washington, USA, http://pubs.wri.org/pubs_pdf.cfm?PubID=3762.

Kindermann, G., M. Obersteiner, B. Sohngen, J. Sathaye, K. Andrasko, E. Rametsteiner, B. Schlamadinger, S. Wunder, and R. Beach. 2008. "Global cost estimates of reducing carbon emissions through avoided deforestation." *Proceedings of the National Academy of Sciences* 105(30): 10302.

Kirchsteiger, C. 2006. "Current practices for risk zoning around nuclear power plants in comparison to other industry sectors." *Journal of Hazardous Materials* 136(3): 392–397.

Kissinger, G. 2011. "Linking forests and food production in the REDD+ context." CCAFS Policy Brief no. 3. CGIAR Research Program on Climate Change, Agriculture and Food Security (CCAFS). Copenhagen, Denmark. Online at www.ccafs.cgiar.org.

Kitchendaily. 2010. "Loopholes in Trading Greenhouse Gases". Kitchendaily.

Kiun, E.H. 1996. "Analyzing and managing policy processes in complex networks." *Administration and Society* 28(1): 90–119.

Klauer, B., M. Drechsler, and F. Messner. 2006. "Multicriteria analysis under uncertainty with IANUS–method and empirical results." *Environment and Planning C: Government and Policy* 24(2): 235–256.

Klein, R.J.T., E.L.F. Schipper, and S. Dessai. 2005. "Integrating mitigation and adaptation into climate and development policy: three research questions." *Environmental Science and Policy* 8(6): 579–588.

Klein, R.W. and S. Wang. 2009. "Catastrophe risk financing in the United States and the European Union: a comparative analysis of alternative regulatory approaches." *Journal of Risk and Insurance* 76(3): 607–637.

Klijn, E.H. 1996. "Analyzing and managing policy processes in complex networks." *Administration and Society* 28: 90–119.

Klijn, E.H. 2001. "Rules as institutional context for decision making in networks." *Administration and Society* 33(2): 133–164.

Klijn, E.H. and C. Skelcher. 2007. "Democracy and governance networks: compatible or not?" *Public Administration* 85(3): 587–608.

Klinsky, S. and H. Dowlatabadi. 2009. "Conceptualizations of justice in climate policy." *Climate Policy* 9(1): 88–108.

Kloosterman, K. 2011. "Watts to Water Brothers in rural Pakistan" Green Prophet.

Koliba, C., J.W. Meek, and A. Zia. 2010. *Governance networks in public administration and public policy*: CRC Press, Inc.

Koliba, C.J., R.M. Mills, and A. Zia. 2011. "Accountability in governance networks: an assessment of public, private, and nonprofit emergency management practices following hurricane Katrina." *Public Administration Review* 71(2): 210–220.

Kooiman, J. 2008. "Exploring the concept of governability." *Journal of Comparative Policy Analysis: Research and Practice* 10(2): 171–190.

Koontz, T.M. 2004. *Collaborative environmental management: what roles for government?* RFF Press.

Koppenjan, J. and E.H. Klijn. 2004. *Managing uncertainties in networks: a network approach to problem solving and decision making*: Routledge.

Kosko, B. 1986. "Fuzzy cognitive maps." *International Journal of Man-Machine Studies* 24(1): 65–75.

Kovacevic, M. 2011. "REDD+ best chance for progress on climate change at Durban, says scientist." *Forest news: a blog* by the Center for International Forestry Research.

Kriesel, W. and C. Landry. 2004. "Participation in the National Flood Insurance Program: an empirical analysis for coastal properties." *Journal of Risk and Insurance* 71(3): 405–420.

Krukowska, E. 2011. "Russia may become third-biggest CO_2 offsets supplier, CDC says." *Bloomberg News*.

Kruppa, M., and Allan, A. 2011. "Carbon trading may be ready for its next act." *New York Times*, November 13, 2011.

Kühberger, A. 1998. "The influence of framing on risky decisions: a meta-analysis." *Organizational Behavior and Human Decision Processes* 75(1): 23–55.

Kunreuther, H. 2008. "Reducing losses from catastrophic risks through long-term insurance and mitigation." *Social Research: An International Quarterly* 75(3): 905–930.

Kurbanov, E., O. Vorobyov, A. Gubayev, L. Moshkina, and S. Lezhnin. 2007. "Carbon sequestration after pine afforestation on marginal lands in the Povolgie region of Russia: a case study of the potential for a joint implementation activity." *Scandinavian Journal of Forest Research* 22(6): 488–499.

Kwon, T.H. 2005. "Decomposition of factors determining the trend of CO_2 emissions from car travel in Great Britain (1970–2000)." *Ecological Economics* 53(2): 261–275.

Kyoto Protocol. 1997. "United Nations framework convention on climate change." Kyoto Protocol.

Kyoto Protocol, The United Nations Framework Convention on Climate Change (15th Conference of the Parties – The Copenhagen). 2009. "Background paper A: setting the targets." The United Nations Framework Convention on Climate Change (15th Conference of the Parties – The Copenhagen Protocol).

Lang, C. 2009. "How the World Bank explains REDD to indigenous peoples." Redd-monitor.org.

Lang, C. 2010. "Environmental defense fund caught REDD-handed." Redd-monitor.org.

Langholz, J. 1999. "Exploring the effects of alternative income opportunities on rainforest use: insights from Guatemala's Maya Biosphere Reserve." *Society and Natural Resources* 12(2): 139–149.

Larson, A.M. 2011. "Forest tenure reform in the age of climate change: lessons for REDD+." *Global Environmental Change* 21(2): 540–549.

Larson, A.M. and E. Petkova. 2011. "An introduction to forest governance, people and REDD+ in Latin America: obstacles and opportunities." *Forests* 2(1): 86–111.

Laukkanen, S., A. Kangas, and J. Kangas. 2002. "Applying voting theory in natural resource management: a case of multiple-criteria group decision support." *Journal of Environmental Management* 64(2): 127–137.

Lebel, L., P. Garden, and M. Imamura. 2005. "The politics of scale, position, and place in the governance of water resources in the Mekong region." *Ecology and Society* 10(2): 18.

Lederer, M. 2011. "From CDM to REDD+: what do we know for setting up effective and legitimate carbon governance?" *Ecological Economics*.

Ledgerwood, A., C.J. Wakslak, and M.A. Wang. 2010. "Differential information use for near and distant decisions." *Journal of Experimental Social Psychology* 40: 638–642.

Leimbach, M. 2003. "Equity and carbon emissions trading: a model analysis." *Energy Policy* 31(10): 1033–1044.

Leiserowitz, A. 2006. "Climate change risk perception and policy preferences: the role of affect, imagery, and values." *Climatic Change* 77(1): 45–72.

Leiserowitz, A. 2007. "Communicating the risks of global warming: American risk perceptions, affective images, and interpretive communities." *Creating a Climate For Change: Communicating Climate Change and Facilitating Social Change*: 44–63.

Leiss, W. 1995. "Down and dirty: the use and abuse of public trust in risk communication." *Risk Analysis* 15(6): 685–692.

Levy, B. 2007. "Stemming the tide of the uninsured." *Policy Studies Journal* 35: 554.

Linnerooth-Bayer, J. and R. Mechler. 2007. "Disaster safety nets for developing countries: extending public–private partnerships." *Environmental Hazards* 7(1): 54–61.

Linnerooth-Bayer, J. and R. Mechler. 2009. "Insurance against losses from natural disasters in developing countries." Working paper. Online at www.microfinancegateway.org/gm/document-1.9.40643/13.pdf.

Linnerooth-Bayer, J., M.J. Mace, and R. Verheyen. 2003. *Insurance-related actions and risk assessment in the context of the UNFCCC*: UNFCCC Secretariat. Online at http://unfccc.int/files/meetings/workshops/other_meetings/application/pdf/background.pdf.

Lise, W. 2006. "Decomposition of CO_2 emissions over 1980–2003 in Turkey." *Energy Policy* 34(14): 1841–1852.

LittleREDDdesk. 2009. "Introduction to REDD." LittleREDDdesk.

Liu, J. *et al.* 2007. "Complexity of coupled human and natural systems." *Science* 317(5844): 1513–1516.

Loewenstein, G. 1996."Out of control: visceral influences on behavior." *Organizational Behavior and Human Decision Processes* 65: 272–292.

Loewenstein, G. 2003. *Time and decision: economic and psychological perspectives on intertemporal choice*: Russell Sage Foundation Publications.

Lorenzoni, I. and N.F. Pidgeon. 2006. "Public views on climate change: European and USA perspectives." *Climatic Change* 77(1): 73–95.

Lovell, H., H. Bulkeley and D. Liverman. 2009. "Carbon offsetting: sustaining consumption?" *Environment and Planning A* 41: 2357–2379.

Lowndes, V. and C. Skelcher. 1998. "The dynamics of multi-organizational partnerships: an analysis of changing modes of governance." *Public Administration* 76(2): 313–333.

Lozano, S. and E. Gutiérrez. 2008. "Non-parametric frontier approach to modelling the relationships among population, GDP, energy consumption and CO_2 emissions." *Ecological Economics* 66(4): 687–699.

Luechinger, S. and P.A. Raschky. 2009. "Valuing flood disasters using the life satisfaction approach." *Journal of Public Economics* 93(3–4): 620–633.

Lund, E. 2010. "Dysfunctional delegation: why the design of the CDM's supervisory system is fundamentally flawed." *Climate Policy* 10(3): 277–288.

Luterbacher, U. and D.F. Sprinz. 2001. *International relations and global climate change*: MIT Press.

Lynn Jr, L.E., C.J. Heinrich and C.J. Hill. 2000. "Studying governance and public management: Challenges and prospects." *Journal of Public Administration Research and Theory* 10(2): 233–262.

Macdonald, A. 2010. "Improving or disproving sustainable development in the clean development mechanism in the midst of a financial crisis?" *Law Environment and Development Journal* 1: 1, 15.

McCarthy, J.J. *et al.* 2001. *Climate change 2001: impacts, adaptation, and vulnerability: contribution of Working Group II to the third assessment report of the Intergovernmental Panel on Climate Change*: Cambridge University Press.

MacKellar, L., P. Freeman, and T. Ermolieva. 1999. "Estimating natural catastrophic risk exposure and the benefits of risk transfer in developing countries." Paper presented at World Bank Meeting "Issues for a Consultative Group for Global Disaster Reduction," Paris, June 1–2, 1999. Available at www.iiasa.ac.at/Research/CAT/paris.pdf.

McKenzie, R. and J. Levendis. 2010. "Flood hazards and urban housing markets: the effects of Katrina on New Orleans." *The Journal of Real Estate Finance and Economics* 40(1): 62–76.

McKibben, B. 2011. *Earth: making a life on a tough new planet*: St. Martin's Griffin.

Maclellan, N. 2011. "Improving access to climate funding for the Pacific Islands." *Development policy blog* from the Development Policy Centre.

MacMynowski, D.P. 2007. "Pausing at the brink of interdisciplinarity: power and knowledge at the meeting of the social and biophysical science." *Ecology and Society* 12(1): 20.

McShane, T. (ed.) 2011. *Advancing conservation in a social context: working in a world of trade-offs*. Global Institute of Sustainability, Arizona State University.

McShane, T.O., P.D. Hirsch, T.C. Trung, A.N. Songorwa, A. Kinzig, B. Monteferri, D. Mutekanga, H.V. Thang, J.L. Dammert, and M. Pulgar-Vidal. 2011. "Hard choices: making trade-offs between biodiversity conservation and human well-being." *Biological Conservation* 144(3): 966–972.

Mardas, N., Mitchell, A., Crosbie, L., Ripley, S., Howard, R., Elia, C., and Trivedi, M. 2009. "Global forest footprints." *Forest Footprint Disclosure Project*: Global Canopy Program.

Marsh, D. and R.A.W. Rhodes. 1992. *Policy networks in British government*: Oxford University Press.

Marston, S.A. 2000. "The social construction of scale." *Progress in Human Geography* 24(2): 219–242.

Martello, R. and P. Dargusch. 2010. "A systems analysis of factors affecting leakage in reduced emissions from deforestation and degradation projects in tropical forests in developing nations." *Small-Scale Forestry* 9(4): 501–516.

Martinez-Alier, J. 2001. "Ecological conflicts and valuation: mangroves versus shrimps in the late 1990s." *Environment and Planning C* 19(5): 713–728.

Martinez-Alier, J., G. Munda, and J. O'Neill. 1998. "Weak comparability of values as a foundation for ecological economics." *Ecological Economics* 26(3): 277–286.

Marttunen, M. and R.P. Hämäläinen. 2008. "The decision analysis interview approach in the collaborative management of a large regulated water course." *Environmental Management* 42(6): 1026–1042.

Mashaw, J.L. 2006. "Accountability and institutional design: some thoughts on the grammar of governance." *Public Accountability: Designs, Dilemmas and Experiences*: 115–156.

Mathur, N. and C. Skelcher. 2007. "Evaluating democratic performance: methodologies for assessing the relationship between network governance and citizens." *Public Administration Review* 67(2): 228–237.

May, P.J. 2007. "Regulatory regimes and accountability." *Regulation and Governance* 1(1): 8–26.

Meenar, M.M., J. Duffy, and ASM Bari. 2006. "Life on the floodplain: remapping watersheds, neighborhoods, and lives." *Planning-Chicago* 72(7): 30.

Messner, F. 2006. "Applying participatory multicriteria methods to river basin management: improving the implementation of the Water Framework Directive." *Environment and Planning C: Government and Policy* 24(2): 159–167.

Messner, F., O. Zwirner, and M. Karkuschke. 2006. "Participation in multi-criteria decision support for the resolution of a water allocation problem in the Spree River basin." *Land Use Policy* 23(1): 63–75.

METI. 2003. "Perspectives and actions to construct a future sustainable framework on climate change." Interim Report by the Global Environmental Subcommittee, Environmental Committee, Industrial Structure Council, METI.

Meyer, A. 2000. *Contraction and convergence: the global solution to climate change*: Green Books.

Michaelowa, A. and H. Schmidt. 1997. "A dynamic crediting regime for joint implementation to foster innovation in the long term." *Mitigation and Adaptation Strategies for Global Change* 2(1): 45–56.

Michaelowa, A., S. Butzengeiger, and M. Jung. 2005. "Graduation and deepening: an ambitious post-2012 climate policy scenario." *International Environmental Agreements: Politics, Law, and Economics* 5(1): 25–46.

Michel, D. 2004. *Climate policy for the 21st century: meeting the long-term challenge of global warming*: Center for Transatlantic Relations.

Michel, D. 2009. "Foxes, hedgehogs, and greenhouse governance: knowledge, uncertainty, and international policy-making in a warming world." *Applied Energy* 86(2): 258–264.

Milward, H.B. 1996. "Symposium on the hollow state: capacity, control, and performance in interorganizational settings." *Journal of Public Administration Research and Theory* 6(2): 193–195.

Milward, H.B. 1996. "Introduction." *Journal of Public Administration Research and Theory* 6(2): 193–195.

Mingus, M.S. 2004. "Validating the comparative network framework in a Canada/United States context." *Journal of Comparative Policy Analysis: Research and Practice* 6(1): 15–37.

Mintz, A. and C. Huang. 1991. "Guns versus butter: the indirect link." *American Journal of Political Science*: 738–757.

Mohapatra, P.K., and R.D. Singh. 2003. "Flood management in India." *Natural Hazards* 28(1): 131–143.

Montz, B. and E.C. Gruntfest. 1986. "Changes in American urban floodplain occupancy since 1958: the experiences of nine cities." *Applied Geography* 6(4): 325–338.

Morgan, A. 2007. "The impact of Hurricane Ivan on expected flood losses, perceived flood risk, and property values." *Journal of Housing Research* 16(1): 47–60.

Morgan, M.G. 2002. *Risk communication: a mental models approach*: Cambridge University Press.

Morgan, M.G., B. Fischoff, A. Bostrom, and C.J. Atman. 2002. *Risk communication: a mental models approach*: Cambridge University Press.

Moser, S.C. and L. Dilling. 2004. "Making climate hot." *Environment: Science and Policy for Sustainable Development* 46(10): 32–46.

Moser, S.C. and L. Dilling. 2007. *Creating a climate for change: communicating climate change and facilitating social change*: Cambridge University Press.

Moynihan, D.P. 2008. "Learning under uncertainty: networks in crisis management." *Public Administration Review* 68(2): 350–365.

Muckleston, K.W. 1983. "The impact of floodplain regulations on residential land values in Oregon." *Journal of the American Water Resources Association* 19(1): 1–7.

Müller, B. 1999. "Justice in global warming negotiations: how to obtain a procedurally fair compromise." Oxford Institute for Energy Studies.

Müller, B. 2001. "Fair compromise in a morally complex world." Oxford Institute for Energy Studies.

Munda, G. 2006. "Social multi-criteria evaluation for urban sustainability policies." *Land Use Policy* 23(1): 86–94.

Murphy, A.H., S. Lichtenstein, B. Fischhoff, and R.L. Winkler. 1980. "Misinterpretations of precipitation probability forecasts." *Bulletin of American Meteorological Society* 61: 695–701.

Nabuurs, G.J., O. Masera, K. Andrasko, P. Benitez-Ponce, R. Boer, M. Dutschke, E. Elsiddig, J. Ford-Robertson, P. Frumhoff, T. Karjalainen, O. Krankina, W.A. Kurz, M. Matsumoto, W. Oyhantcabal, N.H. Ravindranath, M.J. Sanz Sanchez, and X. Zhang, 2007. "Forestry." In *Climate change 2007: mitigation. Contribution of Working Group III to the Fourth Assessment Report of the Intergovernmental Panel on Climate Change*, O.R. Davidson B. Metz, P.R. Bosch, R. Dave, and L.A. Meyer (eds.): Cambridge University Press.

Naidoo, B. 2011. "As the COP 17 clock counts down, few signs of global consensus emerge." *Engineering News*.

Najam, A. and A. Sagar. 1998. "Avoiding a COP-out: moving towards systematic decision-making under the climate convention." *Climatic Change* 39(4).

Najam, A., S. Huq, and Y. Sokona. 2003. "Climate negotiations beyond Kyoto: developing countries concerns and interests." *Climate Policy* 3(3): 221–231.

NSF (National Science Foundation). 1999. "Solving the puzzle: researching the impacts of climate change around the world." NSF: Washington DC. Online at www.nsf.gov/news/special_reports/climate/pdf/NSF_Climate_Change_Report.pdf.

Nazifi, F. 2010. "The price impacts of linking the European Union Emissions Trading Scheme to the Clean Development Mechanism." *Environmental Economics and Policy Studies* 12(4): 164–186.

Neeff, T. and F. Ascui. 2009. "Lessons from carbon markets for designing an effective REDD architecture." *Climate Policy* 9(3): 306–315.

Nepstad, D., F.M. Britaldo, S. Soares-Filho, A. Lima, P. Moutinho, M. Bowman, J. Carter, A. Cattaneo, H. Rodrigues, S. Schwartzman, C.M. Stickler, D.G. McGrath, R. Lubowski, P. Piris-Cabezas, A. Alencar Sergio Rivero, O. Almeida, and O. Stella. 2009. "The end of deforestation in the Brazilian Amazon." *Science* 326: 1350–1351.

Newell, P. and M. Paterson. 2010. *Climate capitalism: global warming and the transformation of the global economy*: Cambridge University Press.

Ninomiya, Y. 2003. "Prospects for energy efficiency improvement through an international agreement." *Climate Regime Beyond 2012: Incentives for Global Participation*: 16–19.

Nordhaus, W.D. 1994. *Managing the global commons: the economics of climate change*: MIT Press.

Norese, M.F. 2006. "ELECTRE III as a support for participatory decision-making on the localisation of waste-treatment plants." *Land Use Policy* 23(1): 76–85.

Norgaard, R.B. 2010. "Ecosystem services: from eye-opening metaphor to complexity blinder." *Ecological Economics* 69(6): 1219–1227.

Norman, C.R. 2011. "Progress slow on global climate mitigation" *Tico Times*, November 18, 2011. Online at www.ticotimes.net/Current-Edition/Top-Story/News/Progress-slow-on-global-climate-mitigation_Friday-November-18-2011.

Norton, B. 1996. "Integration or reduction: two approaches to environmental values." *Environmental Pragmatism*: 105–138.

Norton, B., R. Costanza, and R.C. Bishop. 1998. "The evolution of preferences: why sovereign preferences may not lead to sustainable policies and what to do about it." *Ecological Economics* 24(2): 193–211.

Norton, B.G. 1987. *Why preserve natural variety?* Princeton University Press.

Norton, B.G. 1989. "Intergenerational equity and environmental decisions: a model using Rawls' veil of ignorance." *Ecological Economics* 1(2): 137–159.

Norton, B.G. 1991. "Thoreau's insect analogies: or, why environmentalists hate mainstream economists." *Environmental Ethics* 13(3): 235–251.

Norton, B.G. 1994. "Economists' preferences and the preferences of economists." *Environmental Values* 3(4): 311–332.

Norton, B.G. 2000. "Biodiversity and environmental values: in search of a universal earth ethic." *Biodiversity and Conservation* 9(8): 1029–1044.

Norton, B.G. 2005. *Sustainability: a philosophy of adaptive ecosystem management*: University of Chicago Press.

Norton, B.G. 2007. "Ethics and sustainable development: an adaptive approach to environmental choice." *Handbook of Sustainable Development*: 27.

Norton, B.G. and D. Noonan. 2007. "Ecology and valuation: big changes needed." *Ecological Economics* 63(4): 664–675.

Norton, B.G. and A.C. Steinemann. 2001. "Environmental values and adaptive management." *Environmental Values* 10(4): 473–506.

Norton, B.G. and M.A. Toman. 1997. "Sustainability: ecological and economic perspectives." *Land Economics*: 553–568.

O'Connor, R.E., B. Yarnal, K. Dow, C.L. Jocoy, and G.J. Carbone. 2005. "Feeling at risk matters: water managers and the decision to use forecasts." *Risk Analysis* 25: 1265–1273.

O'Hara, S.U. 1996. "Discursive ethics in ecosystems valuation and environmental policy." *Ecological Economics* 16(2): 95–107.

Okereke, C., H. Bulkeley and H. Schroeder. 2009. "Conceptualizing climate governance beyond the international regime." *Global Environmental Politics* 9(1): 58–78.

Okogu, B.E. 1996. "Joint implementation: an effective instrument of north–south environmental co-operation?" *Energy for Sustainable Development* 2(5): 21–31.

Olsen, J.R. 2006. "Climate change and floodplain management in the United States." *Climatic Change* 76(3): 407–426.

Olsen, K.H. and J. Fenhann. 2008. "Sustainable development benefits of clean development mechanism projects: a new methodology for sustainability assessment based on text analysis of the project design documents submitted for validation." *Energy Policy* 36(8): 2819–2830.

O'Neill, B.C. and B.S. Chen. 2002. "Demographic determinants of household energy use in the United States." *Population and Development Review* 28: 53–88.

Openshaw, S. 1984. *The modifiable areal unit problem, concepts and techniques in modern geography*, series no. 38: Geo Books.

Ostrom, E. 2005. *Understanding institutional diversity*: Princeton University Press.

Ostrom, E. 2007. "A diagnostic approach for going beyond panaceas." *Proceedings of the National Academy of Sciences* 104(39): 15181.

Ott, H.E., H. Winkler, B. Brouns, S. Kartha, M. Mace, S. Huq, Y. Kameyama, A.P. Sari, J. Pan, and Y. Sokona. 2004. "South–north dialogue on equity in the greenhouse." *A proposal for an adequate and equitable global climate agreement*: Eschborn, GTZ.

Padma, T.V. 2011. "CDM emerging a disincentive for government investment in renewables." Science and Development Network.

Page, S. 2004. "Measuring accountability for results in interagency collaboratives." *Public Administration Review* 64(5): 591–606.

Page, T. 1997. "On the problem of achieving efficiency and equity, intergenerationally." *Land Economics*: 580–596.

Palmer, C. 2011. "Property rights and liability for deforestation under REDD+: implications for 'permanence' in policy design." *Ecological Economics* 70(4): 571–576.

Pan, J. 2003. "Commitment to human development goals with low emissions: an alternative to emissions caps for post-Kyoto from a developing country perspective." Presentation at side event hosted by Research Centre for Sustainable Development, Chinese Academy of Social Sciences, UNFCCC COP9 1.

Pan, Y., R.A. Birdsey, Jingyun Fang, R. Houghton, P.E. Kauppi, O.L. Phillips Werner, A. Kurz, A. Shvidenko, S.L. Lewis, J.G. Canadell, R.B. Jackson, P. Ciais, S.W. Pacala, A.D. McGuire, S. Piao, S. Sitch, A. Rautiainen, and D. Hayes. 2010. "A large and persistent carbon sink in the world's forests." *Science* 333: 988–993.

Papadopoulos, Y. 2003. "Cooperative forms of governance: Problems of democratic accountability in complex environments." *European Journal of Political Research* 42(4): 473–501.

Papadopoulos, Y. 2007. "Problems of democratic accountability in network and multilevel governance." *European Law Journal* 13(4): 469–486.

Pardo, R. and F. Calvo. 2002. "Attitudes toward science among the European public: a methodological analysis." *Public Understanding of Science* 11(2): 155–195.

Pardo, R. and F. Calvo. 2004. "The cognitive dimension of public perceptions of science: methodological issues." *Public Understanding of Science* 13(3): 203–227.

Parker, C., A. Mitchell, M. Trivedi, N. Mardas and K. Sosis. 2009. *The little REDD+ book*: Global Canopy Foundation.

Parker, D.J. 1995. "Floodplain development policy in England and Wales." *Applied Geography* 15(4): 341–363.

Parry, M.L., M. Parry, N. Arnell, P. Berry, D. Dodman, S. Fankhauser, C. Hope, S. Kovats, R. Nicholls, and D. Sattherwaite. 2009. *Assessing the costs of adaptation to climate change: a review of the UNFCCC and other recent estimates*: Iied.

Pellizzoni, L. 2001. "The myth of the best argument: power, deliberation and reason." *The British Journal of Sociology* 52(1): 59–86.

Perelman, M. 2000. *The invention of capitalism: classical political economy and the secret history of primitive accumulation*: Duke University Press.

Perelman, M. 2003. "Myths of the market: economics and the environment." *Organization and Environment* 16(2): 199–202.

Peroff, K. and M. Podolak-Warren. 1979. "Does spending on defence cut spending on health? A time-series analysis of the US economy 1929–74." *British Journal of Political Science* 9(1): 21–39.

Perrings, C. and B. Hannon. 2001. "An introduction to spatial discounting." *Journal of Regional Science* 41(1): 23–38.

Perrow, C. 2010. "Why we disagree about climate change: understanding controversy, inaction, and opportunity." *Contemporary Sociology: A Journal of Reviews* 39(1): 46–47.

Pettenger, M.E. 2007. *The social construction of climate change: power, knowledge, norms, discourses*: Ashgate Pub Co.

Phelps, J., E.L. Webb, and L.P. Koh. 2011. "Risky business: an uncertain future for biodiversity conservation finance through REDD+." *Conservation Letters* 4(2): 88–94.

Philibert, C. 2005. "Approaches for future international co-operation." OECD. Online at www.oecd.org/env/climatechange/35009660.pdf.

Pielke Jr, R.A. and M.W. Downton. 2000. "Precipitation and damaging floods: trends in the United States, 1932–97." *Journal of Climate* 13(20): 3625–3637.

Pierce Colfer, C.J. 2011. "Marginalized forest peoples' perceptions of the legitimacy of governance: an exploration." *World Development* 39(12): 2147–2164.

Pierre, J. and G.B. Peters. 2005. *Governing complex societies: trajectories and scenarios*: Palgrave Macmillan.

Pimm, S.L., G.J. Russell, J.L. Gittleman, and T.M. Brooks. 1995. "The future of biodiversity." *Science* 269(5222): 347–350.

Pimm, S.L., M. Ayres, A. Balmford, G. Branch, K. Brandon, T. Brooks, R. Bustamante, R. Costanza, R. Cowling, and L.M. Curran. 2001. "Can we defy nature's end?" *Science* 293(5538): 2207–2208.

Pompe, J.J. and J.R. Rinehart. 2008a. "Property insurance for coastal residents: governments' 'ill wind'." *Independent Review* 13(2): 189–207.

Pompe, J.J. and J.R. Rinehart. 2008b. "Mitigating damage costs from hurricane strikes along the southeastern US Coast: a role for insurance markets." *Ocean and Coastal Management* 51(12): 782–788.

Posner, P.L. 2002. "Accountability challenges of third party governance." *The Tools of Government*: Oxford University Press.

Power, F.B. and E.W. Shows. 1979. "A status report on the National Flood Insurance Program: mid 1978." *Journal of Risk and Insurance*: 61–76.

Power, F.B. and E.W. Shows. 1981. "Failure of the national flood insurance partnership." *Journal of Business Research* 9(4): 397–407.

Price, C. 1993. *Time, discounting and value*: Blackwell Publishers.

Price, L., L. Michaelis, E. Worrell, and M. Khrushch. 1998. "Sectoral trends and driving forces of global energy use and greenhouse gas emissions." *Mitigation and Adaptation Strategies for Global Change* 3(2): 263–319.

Priest, S.H., H. Bonfadelli, and M. Rusanen. 2003. "The 'trust gap' hypothesis: predicting support for biotechnology across national cultures as a function of trust in actors." *Risk Analysis* 23(4): 751–766.

Proctor, W. and M. Dreschler. 2006. "Deliberative multicriteria evaluation." *Environment and Planning C: Government and Policy* 24(2): 169–190.

Provan, K.G. and H.B. Milward. 1995. "A preliminary theory of interorganizational network effectiveness: a comparative study of four community mental health systems." *Administrative science quarterly*: 1–33.

Radin, B. 2006. *Challenging the performance movement: accountability, complexity, and democratic values*: Georgetown University Press.

Raiffa, H. 1968. *Decision analysis: introductory lectures on choices under uncertainty*: Addison-Wesley.

Ramanathan, R. 2001. "A note on the use of the analytic hierarchy process for environmental impact assessment." *Journal of Environmental Management* 63(1): 27–35.

Ramanathan, R. 2006. "A multi-factor efficiency perspective to the relationships among world GDP, energy consumption and carbon dioxide emissions." *Technological Forecasting and Social Change* 73(5): 483–494.

Ramankutty, N., H.K. Gibbs, F. Achard, R. Defries, J.A. Foley, and RA Houghton. 2007. "Challenges to estimating carbon emissions from tropical deforestation." *Global Change Biology* 13(1): 51–66.

RAPID, Europa Press Releases. 2003. "Kyoto Protocol: What is the Kyoto Protocol?" Europa.

Raschky, P.A. and H. Weck-Hannemann. 2007. "Charity hazard: a real hazard to natural disaster insurance?" *Environmental Hazards* 7(4): 321–329.

Raupach, M.R., G. Marland, P. Ciais, C. Le Quéré, J.G. Canadell, G. Klepper, and C.B. Field. 2007. "Global and regional drivers of accelerating CO_2 emissions." *Proceedings of the National Academy of Sciences* 104(24): 10288.

Reinstein, R.A. 2004. "A possible way forward on climate change." *Mitigation and Adaptation Strategies for Global Change* 9(3): 245–309.

Renn, O. 2006. "Participatory processes for designing environmental policies." *Land Use Policy* 23(1): 34–43.

Repetto, R. 2001. "The Clean Development Mechanism: institutional breakthrough or institutional nightmare?" *Policy Sciences* 34(3): 303–327.

Rice, J.L. 2010. "Climate, carbon, territory: greenhouse gas mitigation in Seattle, Washington." *Annals of the Association of American Geographers* 100(4): 929–937

Richman, S. 1993. *Federal flood insurance: managing risk or creating it*: National Emergency Training Center.

Rittel, H.W.J. and M.M. Webber. 1973. "Dilemmas in a general theory of planning." *Policy Sciences* 4(2): 155–169.

Rittel, H.W.J. and M.M. Webber. 1984. "Planning problems are wicked problems." *Developments in Design Methodology*: 135–144.

Rive, N., A. Torvanger, and J.S. Fuglestvedt. 2006. "Climate agreements based on responsibility for global warming: periodic updating, policy choices, and regional costs." *Global Environmental Change* 16(2): 182–194.

Roberts, N. 2000. "Wicked problems and network approaches to resolution." *International Public Management Review* 1(1): 1–19.

Rodríguez, J.P., T.D. Beard, E.M. Bennett, G.S. Cumming, S.J. Cork, J. Agard, A.P. Dobson, and G.D. Peterson. 2006. "Trade-offs across space, time, and ecosystem services." *Ecology and Society* 11(1): 28.

Rodríguez, C., A. Langley, F. Béland, and J.L. Denis. 2007. "Governance, power, and mandated collaboration in an interorganizational network." *Administration and Society* 39(2): 150–193.

Romm, J.J. 2007. *Hell and high water: global warming-the solution and the politics-and what we should do*: William Morrow and Co.

Rosenthal, E. 2009. "Climate Talks Near Deal to Save Forests." *The New York Times*, December 15, 2009.

Rübbelke, D.T.G. 2011. "International support of climate change policies in developing countries: strategic, moral and fairness aspects." *Ecological Economics*.

Rudolf, John C. 2010. "On our radar: skepticism on carbon offsets." *Green: a blog about energy and the environment*: *The New York Times*.

Russett, B.M. 1969. "Who pays for defense?" *The American Political Science Review* 63(2): 412–426.

Russett, B.M. 1970. *What price vigilance? The burdens of national defense*: Yale University Press New Haven.

Sabatier, P.A. 2007. *Theories of the policy process*: Westview Press.

Sagoff, M. 1998. "Aggregation and deliberation in valuing environmental public goods: a look beyond contingent pricing." *Ecological Economics* 24(2): 213–230.

Samuelson, P.A. 1937. "A note on measurement of utility." *The Review of Economic Studies* 4(2): 155–161.

Sanyal, J. and XX L. 2004. "Application of remote sensing in flood management with special reference to monsoon Asia: a review." *Natural Hazards* 33(2): 283–301.

Sanyal, J. and XX L. 2006. "GIS-based flood hazard mapping at different administrative scales: a case study in Gangetic West Bengal, India." *Singapore Journal of Tropical Geography* 27(2): 207–220.

Sarewitz, D., R. Pielke Jr, and M. Keykhah. 2003. "Vulnerability and risk: some thoughts from a political and policy perspective." *Risk Analysis* 23(4): 805–810.

Sawyer, Steve. 2011. "Comment: Cancun – a modest success?" *Renewable Energy Focus*.

Scheffer, M. 2009. *Critical transitions in nature and society*: Princeton University Press.

Schelling, T.C. 1997. "The cost of combating global warming: facing the tradeoffs." *Foreign Affairs*: 8–14.

Schelling, T.C. 2002. "What makes greenhouse sense? Time to rethink the Kyoto Protocol." *Foreign Affairs*: 2–9.

Schiermeier, Q. 2011. "Clean-energy credits tarnished." *Nature* 477(7366): 517.

Schmalensee, R. 1996. *Greenhouse policy architecture and institutions*: MIT Joint Program on the Science and Policy of Climate Change. Online at http://web.mit.edu/globalchange/www/MITJPSPGC_Rpt13.pdf. Retrieved March 10, 2009.

Schmalensee, R. 1998. "Greenhouse policy architecture and institutions." *Economics and Policy Issues in Climate Change*: 137–158.

Schmidt, J., N. Helme, J. Lee, and M. Houdashelt. 2008. "Sector-based approach to the post-2012 climate change policy architecture." *Climate Policy* 8(5): 494–515.

Schneider, L. 2009. "A Clean Development Mechanism with global atmospheric benefits for a post-2012 climate regime." *International Environmental Agreements: Politics, Law and Economics* 9(2): 95–111.

Schroeder, R.A. 1999. "Geographies of environmental intervention in Africa." *Progress in Human Geography* 23(3): 359–378.

Scott, C. 2006. "Spontaneous accountability." *Public Accountability: Designs, Dilemmas and Experiences*: 174–194.

Seymour, F. 2008. "Forests, climate change, and human rights: managing risk and trade-offs." *Human rights and climate change*: Cambridge University Press.

Shapley, L. 1953. "A value for n-person games." In H.W. Kuhn and A.W. Tucker (eds.) *Contributions to the theory of games*: Princeton University.

Sharma, K.P., N.R. Adhikari, P.K. Ghimire, and P.S. Chapagain. 2006. "GIS-based flood risk zoning of the Khando river basin in the Terai region of east Nepal." *Himalayan Journal of Sciences* 1(2): 103–106.

Sharma, S. 2011. "Global warming, CDM and coal power plants." *Counter Currents*, August, 7.

Shilling, J.D., C.F. Sirmans, and J.D. Benjamin. 1989. "Flood insurance, wealth redistribution, and urban property values." *Journal of Urban Economics* 26(1): 43–53.

Shim, JP, M. Warkentin, J.F. Courtney, D.J. Power, R. Sharda, and C. Carlsson. 2002. "Past, present, and future of decision support technology." *Decision Support Systems* 33(2): 111–126.

Shrubsole, D. and J. Scherer. 1996. "Floodplain regulation and the perceptions of the real estate sector in Brantford and Cambridge, Ontario, Canada." *Geoforum* 27(4): 509–525.

Sijm, J., J. Jansen, and A. Torvanger. 2001. "Differentiation of mitigation commitments: the multi-sector convergence approach." *Climate Policy* 1(4): 481–497.

Silver, J.J. 2008. "Weighing in on scale: synthesizing disciplinary approaches to scale in the context of building interdisciplinary resource management." *Society and Natural Resources* 21(10): 921–929.

Simon, H.A. 1982. *Models of Bounded Rationality*: MIT Press.

Skelcher, C. 2005. "Jurisdictional integrity, polycentrism, and the design of democratic governance." *Governance* 18(1): 89–110.

Slaughter, A.M. 2004. "Disaggregated sovereignty: towards the public accountability of global government networks." *Government and Opposition* 39(2): 159–190.

Sloman, S.A. 1996. "The empirical case for two systems of reasoning." Psychological Bulletin 1(119): 3–22.

Slovic, P.E. 2000. *The perception of risk*: Earthscan Publications.

Slovic, P., M.L. Finucane, E. Peters, and D.G. MacGregor. 2004. "Risk as analysis and risk as feelings: some thoughts about affect, reason, risk, and rationality." *Risk Analysis* 24 (2): 311–322.

Smith, J.B. and S.S. Lenhart. 1996. "Climate change adaptation policy options." *Climate Research* 6: 193–201.

Smith, N. 1992. "Geography, difference and the politics of scale." *Postmodernism and the Social Sciences*: 57–79.

Smith, N. 1993. "Homeless/global: scaling places." In J. Bird,(ed.), *Mapping the futures: local cultures global change*, 87–119: Routledge.

Smith, N. 1995. "Remaking scale: competition and cooperation in pre-national and post-national Europe." *State/Space*: 225–238.

Smith, N., J. Bird, B. Curtis, T. Putnam, G. Robertson, and L. Tickner. 1993. *Mapping the Futures: Local Cultures, Global Change*: Routledge.

Solow, R.M. 1993. "Sustainability: an economist's perspective." *Economics of the Environment: Selected Readings* 3: 179–187.

Songorwa, A.N. 1999. "Community-based wildlife management (CWM) in Tanzania: are the communities interested?" *World \Development* 27(12): 2061–2079.

Sørensen, E. and J. Torfing. 2005. "Network governance and post-liberal democracy." *Administrative Theory and Praxis* 27(2): 197–237.

Spash, C.L. 2008. "How much is that ecosystem in the window? The one with the biodiverse trail." *Environmental Values* 17(2): 259–284.

Spash, C.L. and A. Vatn. 2006. "Transferring environmental value estimates: issues and alternatives." *Ecological Economics* 60(2): 379–388.

Spence, A. and N. Pidgeon 2010. "Framing and communicating climate change: the effects of distance and outcome frame manipulations." *Global Environmental Change* 20(4): 656–667.

Stagl, S. 2006. "Multicriteria evaluation and public participation: the case of UK energy policy." *Land Use Policy* 23(1): 53–62.

Steiner, A. 2011. "Global visions of a low-carbon future." *Telegraph*, November 25, 2011. Online at www.telegraph.co.uk/sponsored/earth/the-age-of-energy/8915886/global-views-low-carbon-future.html.

Sterman, J.D. and L.B. Sweeney. 2007. "Understanding public complacency about climate change: adults' mental models of climate change violate conservation of matter." *Climatic Change* 80(3): 213–238.

Stern, N. 2008. "The economics of climate change." *The American Economic Review* 98(2): 1–37.

Stevens, M.R., P.R. Berke, and Y. Song. 2010. "Creating disaster-resilient communities: evaluating the promise and performance of new urbanism." *Landscape and Urban Planning* 94(2): 105–115.

Stevens, W.K. 1997. "Greenhouse gas issue: haggling over fairness." *New York Times*, November 30, 1997. Online at www.nytimes.com/1997/11/30/world/greenhouse-gas-issue-haggling-over-fairness.html?pagewanted=all&src=pm.

Stevens, W.K. 1998. "Pressure builds for accord at talks on climate change." *New York Times*, November 14, 1998. Online at www.nytimes.com/1998/11/14/us/pressure-builds-for-accord-at-talks-on-climate-change.html.

Stewart, R.B. and J.B. Wiener. 2002. *Reconstructing climate policy*: AEI Press.

Sitch, S., B. Smith, I.C. Prentice, A. Arneth, A. Bondeau, W. Cramer, J.O. Kaplan, S. Levis, W. Lucht, M.T. Sykes, K. Thonicke, and S. Venevsky. 2003. "Evaluation of ecosystem dynamics, plant geography and terrestrial carbon cycling in the LPJ Dynamic Global Vegetation Model." *Global Change Biology* 9: 161–185.

Stickler, C.M., D.C. Nepstad, M.T. Coe, D.G. McGrath, H.O. Rodrigues, W.S. Walker, B.S. Soares-Filho, and E.A. Davidson. 2009. "The potential ecological costs and cobenefits of REDD: a critical review and case study from the Amazon region." *Global Change Biology* 15(12): 2803–2824.

Stirling, A. 2006. "Analysis, participation and power: justification and closure in participatory multi-criteria analysis." *Land Use Policy* 23(1): 95–107.

Stoll-Kleemann, S., T. O'Riordan, and C.C. Jaeger. 2001. "The psychology of denial concerning climate mitigation measures: evidence from Swiss focus groups." *Global Environmental Change* 11(2): 107–117.

Strassburg, B., R. Kerry Turner, B. Fisher, R. Schaeffer, and A. Lovett. 2009. "Reducing emissions from deforestation: the 'combined incentives' mechanism and empirical simulations." *Global Environmental Change* 19: 265–278.

Subbarao, S., and B. Lloyd. 2011. "Can the Clean Development Mechanism (CDM) deliver?" *Energy Policy.*

Sugiyama, T. 2003. "Orchestra of treaties: scenario for after 2012." PowerPoint presentation made to UNFCCC in a side event. Available at www.powershow.com/view1/273059-YjQyM/Orchestra_of_Treaties_Scenario_for_after_2012_powerpoint_ppt_presentation.

Sugiyama, T. and J. Sinton. 2005. "Orchestra of treaties: a future climate regime scenario with multiple treaties among like-minded countries." *International Environmental Agreements: Politics, Law and Economics* 5(1): 65–88.

Sugiyama, T., K. Tangen, A. Michaelowa, J. Pan, and H. Hasselknippe. 2003. "Scenarios for the global climate regime." Briefing paper. Online at www.fni.no/post2012/briefing_paper.pdf.

Sunderland, T.C.H., C. Ehringhaus, and B.M. Campbell. 2007. "Conservation and development in tropical forest landscapes: a time to face the trade-offs?" *Environmental Conservation* 34(4): 276–279.

Sunstein, C.R. (2006). "The availability heuristic, intuitive cost-benefit analysis, and climate change." *Climatic Change* 77: 195–210.

Sutter, C. and J.C. Parreño. 2007. "Does the current Clean Development Mechanism (CDM) deliver its sustainable development claim? An analysis of officially registered CDM projects." *Climatic Change* 84(1): 75–90.

Svenson, O. 1975. "A unifying interpretation of different models for the integration of information when evaluating gambles." *Scandinavian Journal of Psychology* 16: 187–192.

Swingland, I.R. and Royal Society. 2003. *Carbon, biodiversity, conservation and income: an analysis of a free-market approach to land-use change and forestry in developing and developed countries: papers of a theme*: Royal Society.

Swyngedouw, E. 1997a. "Excluding the other: the production of scale and scaled politics." *Geographies of Economies*: 167–176.

Swyngedouw, E. 1997b. "Power, nature, and the city: the conquest of water and the political ecology of urbanization in Guayaquil, Ecuador: 1880–1990." *Environment and Planning A* 29: 311–332.

Swyngedouw, E. 1997c. "Neither global nor local: 'glocalization' and the politics of scale." *Reasserting the Power of the Local*: 137–166.

Swyngedouw, E. 2000. "Authoritarian governance, power, and the politics of rescaling." *Environment and Planning D* 18(1): 63–76.

Tan, Celine. 2008. "World Bank's climate funds may undermine UNFCCC talks." *Third World Network*, March 18, 2008. Online at www.twnside.org.sg/title2/finance/twninfofinance20080303.htm.

Tangen, K. and H. Hasselknippe. 2005. "Converging markets." *International Environmental Agreements: Politics, Law and Economics* 5(1): 47–64.

Taylor, S.E. and S.T. Fiske. 1981. "Getting inside the head: methodologies for process analysis in attribution and social cognition." *New Directions in Attribution Research* 3: 459–524.

Teräväinen, T. 2009. "The challenge of sustainability in the politics of climate change: a Finnish perspective on the Clean Development Mechanism." *Politics* 29(3): 173–182.

The Copenhagen Diagnosis. 2009. *Updating the world on latest climate science.* I. Allison, N.L. Bindoff, R.A. Bindschadler, P.M. Cox, N. de Noblet, M.H. England, J.E. Francis, N. Gruber, A.M. Haywood, D.J. Karoly, G. Kaser, C. Le Quéré, T.M. Lenton, M.E. Mann, B.I. McNeil, A.J. Pitman, S. Rahmstorf, E. Rignot, H.J. Schellnhuber,

S.H. Schneider, S.C. Sherwood, R.C.J. Somerville, K. Steffen, E.J. Steig, M. Visbeck, A.J. Weaver. The University of New South Wales Climate Change Research Centre (CCRC), Sydney, Australia.

Thies, C. and R. Czebiniak. 2008. "Forests for climate: developing a hybrid approach for REDD." Greenpeace International.

Thompson, M.C., M. Baruah, and E.R. Carr. 2011. "Seeing REDD+ as a project of environmental governance." *Environmental Science and Policy.*

Tjell, J.C. 2006. "CDM and JI in SWM: clean development mechanisms and joint implementation in solid waste management." *Waste Management and Research* 24(3): 195.

Tobin, G.A. 1999. "Sustainability and community resilience: the holy grail of hazards planning?" *Global Environmental Change Part B: Environmental Hazards* 1(1): 13–25.

Tobin, G.A. and B.E. Montz. 1988. "Catastrophic flooding and the response of the real estate market." *The Social Science Journal* 25(2): 167–177.

Tol, R.S.J. 2007. "The double trade-off between adaptation and mitigation for sea level rise: an application of FUND." *Mitigation and Adaptation Strategies for Global Change* 12(5): 741–753.

Toman, M.A. 1994. "Economics and sustainability: balancing trade-offs and imperatives." *Land Economics*: 399–413.

Torfing, J. 2005. "Governance network theory: towards a second generation." *European Political Science* 4(3): 305–315.

Torvanger, A. and L. Ringius. 2002. "Criteria for evaluation of burden-sharing rules in international climate policy." *International Environmental Agreements: Politics, Law and Economics* 2(3): 221–235.

Tralli, D.M., R.G. Blom, V. Zlotnicki, A. Donnellan, and D.L. Evans. 2005. "Satellite remote sensing of earthquake, volcano, flood, landslide and coastal inundation hazards." *ISPRS Journal of Photogrammetry and Remote Sensing* 59(4): 185–198.

Troy, A. and J. Romm. 2004. "Assessing the price effects of flood hazard disclosure under the California natural hazard disclosure law (AB 1195)." *Journal of Environmental Planning and Management* 47(1): 137–162.

Trumbo, C. 1996. "Constructing climate change: claims and frames in US news coverage of an environmental issue." *Public Understanding of Science* 5(3): 269–283.

Tucker, A.W. 1959. *Contributions to the theory of games*: Princeton University Press.

Tversky, A. and D. Kahneman. 1983. "Extensional vs. intuitive reasoning: the conjunction fallacy in probability judgment." *Psychological Review* 90: 293–315.

Tversky, A. and D. Kahneman. 1986. "Rational choice and the framing of decisions." *Journal of Business* 59(4): 251–278.

Ulrich, W. 1998. "Systems thinking as if people mattered: critical systems thinking for citizens and managers." Working Paper No. 23, Lincoln School of Management, University of Lincolnshire and Humberside.

Ungar, S. 2007. "Public scares: changing the issue culture." *Change* 11: 103–120.

UN-REDD Programme. 2009a. "About REDD+." UN-REDD Programme.

UN-REDD Programme. 2009b. "Multiple benefits: issues and options for REDD." UN-REDD Programme.

United Nations. "Report of the World Commission on Environment and Development: our common future." United Nations.

United Nations. 2010. "Climate change financing will benefit all, UN chief tells Cancun meeting." UN News Centre, United Nations.

United Nations. 2011. "Deforestation threatens planet, economies and communities, UN chief warns." United Nations News Centre, United Nations.

UNFCCC (United Nations Framework Convention on Climate Change). 1998. *Kyoto Protocol to the United Nations Framework Convention on Climate Change*. Online at http://unfccc.int/kyoto_protocol/items/2830.php. Retrieved March 10, 2009.

UNFCCC 2007. "Report of the Conference of the Parties on its Thirteenth Session." Conference of the Parties on its Thirteenth Session, Bali. United Nations.

UNFCCC. 2011. "Clean Development Mechanism." United Nations.

U.S. Department of State. 2002. "US Climate Action Report 2002: third national communication of the United States of America under the United Nations Framework Convention on Climate Change." U.S. Global Change Research Information Office.

Van Dam, C. 2011. "Indigenous territories and REDD in Latin America: opportunity or threat?" *Forests* 2(1): 394–414.

Van Den Hove, S. 2006. "Between consensus and compromise: acknowledging the negotiation dimension in participatory approaches." *Land Use Policy* 23(1): 10–17.

Van der Werf, GR, D.C. Morton, R.S. DeFries, J.G.J. Olivier, P.S. Kasibhatla, R.B. Jackson, G.J. Collatz, and J.T. Randerson. 2009. "CO_2 emissions from forest loss." *Nature Geoscience* 2(11): 737–738.

Vatn, A. 2002. "Efficient or fair: ethical paradoxes in environmental policy." *Economics, Ethics, and Environmental Policy*: 148–163.

Vidal, John. 2009. "Rich nations failing to meet climate aid pledges: world's richest countries have pledged nearly \$18bn to help poorer countries adapt to climate change, but less than \$1bn has been disbursed." *Guardian*, February 20, 2009. Online at www.guardian.co.uk/environment/2009/feb/20/climate-funds-developing-nations.

Vining, A.R., A.E. Boardman, and F. Poschmann. 2005. "Public–private partnerships in the US and Canada: there are no free lunches." *Journal of Comparative Policy Analysis: Research and Practice* 7(3): 199–220.

Wallsten, T.S., S. Fillenbaum, and J.A. Cox. 1986. "Base rate effects on the interpretation of probability and frequency expressions." *Journal of Memory and Language* 25: 571–587.

Wang, C., J. Chen, and J. Zou. 2005. "Decomposition of energy-related CO_2 emission in China: 1957–2000." *Energy* 30(1): 73–83.

WCED (World Commission on Environment and Development). 1987 *Our common future*: United Nations General Assembly.

Weber, E.U. and D.J. Hilton. 1990. "Contextual effects in the interpretations of probability words: perceived base rate and severity of events." *Journal of Experimental Psychology: Human Perception and Performance* 16: 781–789.

Weber, E.P. and A.M. Khademian. 2008. "Wicked problems, knowledge challenges, and collaborative capacity builders in network settings." *Public Administration Review* 68(2): 334–349.

Wei, F., K. Hu, J.L. Lopez, and P. Cui. 2003. "Method and its application of the momentum model for debris flow risk zoning." *Chinese Science Bulletin* 48(6): 594–598.

Weingart, P., A. Engels, and P. Pansegrau. 2000. "Risks of communication: discourses on climate change in science, politics, and the mass media." *Public Understanding of Science* 9(3): 261–283.

Weiss, B.D., E.W. Kligman, and R.L. Reed. 1995. "Literacy skills and communication methods of low-income older persons." *Patient Education and Counseling* 25(2): 109–119.

Wells, M. 1992. "Biodiversity conservation, affluence and poverty: mismatched costs and benefits and efforts to remedy them." *Ambio: A Journal of the Human Environment* 21(3): 237–243.

Wells, M.P. and T.O. McShane. 2004. "Integrating protected area management with local needs and aspirations." *Ambio: A Journal of the Human Environment* 33(8): 513–519.

Wells, P. and G. Paoli. 2011. "An analysis of presidential instruction no. 10, 2011: moratorium on granting of new licenses and improvement of natural primary forest and peatland governance." Daemeter Consulting.

WhatProductions UK. 2008. "Dirty trade: the Clean Development Mechanism (Kevin Smith)." WhatProductions UK.

White, R.D. 2004. *Controversies in environmental sociology*: Cambridge University Press.

Wilensky, H.L. 1974. *The welfare state and equality: structural and ideological roots of public expenditures*: Univ of California Press.

Wilkie, D.S., K.H. Redford, and T.O. McShane. 2010. "Taking of rights for natural resource conservation: a discussion about compensation." *Journal of Sustainable Forestry* 29(2–4): 135–151.

Willis, K.J. and S.A. Bhagwat. 2009. "Biodiversity and Climate Change." *Science* 326: 806–807.

Wilson, E.O. 1989. "Threats to biodiversity." *Scientific American* 261(3): 108–116.

Wilson, E.O. 2002. *The future of life*: Knopf.

Wilson, M.A. and R.B. Howarth. 2002. "Discourse-based valuation of ecosystem services: establishing fair outcomes through group deliberation." *Ecological Economics* 41(3): 431–443.

Winkler, H., R. Spalding-Fecher, S. Mwakasonda, and O. Davidson. 2002. "Sustainable development policies and measures." *Options for Protecting the Climate*, World Resource Institute, Washington DC.

Wittmer, H., F. Rauschmayer, and B. Klauer. 2006. "How to select instruments for the resolution of environmental conflicts?" *Land Use Policy* 23(1): 1–9.

Wood, B.D. and A. Vedlitz. 2007. "Issue definition, information processing, and the politics of global warming." *American Journal of Political Science* 51(3): 552–568.

Wooldridge, J.M. 2000. *Introductory econometrics: a modern approach*: South-Western College Publishing.

Yang, C. and S.H. Schneider. 1997. "Global carbon dioxide emissions scenarios: sensitivity to social and technological factors in three regions." *Mitigation and Adaptation Strategies for Global Change* 2(4): 373–404.

Yanow, D. 1999. *Conducting interpretive policy analysis*: Sage.

Yanow, D. 2000. *Conducting interpretive policy analysis*: Sage Publications, Inc.

Young, A.T. 2008. "Replacing incomplete markets with a complete mess: Katrina and the NFIP." *International Journal of Social Economics* 35(8): 561–568.

Young, O. 1999. *The effectiveness of international environmental regimes: causal connections and behavioral mechanisms. Global environmental accord: strategies for sustainability and institutional innovation*: MIT Press.

Young, O. 2002. *The institutional dimensions of environmental change: fit, interplay, and scale. Global environmental accord: strategies for sustainability and institutional innovation*: The MIT Press.

Zadek, S. 2011. "Beyond climate finance: from accountability to productivity in addressing the climate challenge." *Climate Policy* 11(3): 1058–1068.

Zahabu, E., M.M. Skutsch, H. Sosovele, and R.E. Malimbwi. 2007. "Reduced emissions from deforestation and degradation." *African Journal of Ecology* 45(4): 451–453.

Zahran, S., S. Weiler, S.D. Brody, M.K. Lindell, and W.E. Highfield. 2009. "Modeling national flood insurance policy holding at the county scale in Florida, 1999–2005." *Ecological Economics* 68(10): 2627–2636.

Zaltman, G. 1997. "Rethinking market research: putting people back in." *Journal of Marketing Research*: 424–437.

Zandvoort, H. 2008. "Risk zoning and risk decision making." *International Journal of Risk Assessment and Management* 8(1): 3–18.

Zhang, J. and C. Wang. 2011. "Co-benefits and additionality of the clean development mechanism: an empirical analysis." *Journal of Environmental Economics and Management* 62(2): 140–154.

Zhang, J.X. and C.F. Huang. 2005. "Study on pattern of soft risk zoning map of natural disasters." *Ziran Zaihai Xuebao* 14(6): 20–25.

Zhang, M., H. Mu, and Y. Ning. 2009. "Accounting for energy-related CO_2 emission in China, 1991–2006." *Energy Policy* 37(3): 767–773.

Zhang, Z. 2000. "Decoupling China's carbon emissions increase from economic growth: an economic analysis and policy implications." *World Development* 28(4): 739–752.

Zhouri, A. 2010. "'Adverse forces' in the Brazilian Amazon: developmentalism versus environmentalism and indigenous rights." *The Journal of Environment and Development* 19(3): 252–273.

Zia, A. 1999. *Can beggars be choosers: a critique of Pak-German developmental cooperation*: Heinrich Böll Foundation.

Zia, A. 2004. "Cooperative and non-cooperative decision behaviors in response to the inspection and maintenance program in the Atlanta Airshed, 1997–2001." Ph.D. dissertation, Georgia Institute of Technology. Online at http://smartech.gatech.edu/bitstream/handle/1853/5085/zia_asim__200407_phd.pdf.txt?sequence=2.

Zia, A. 2008. "An ambit-based activity model for evaluating green house gas emission reduction policies." Technical report, Mineta Transportation Institute, San Jose CA. Online at http://transweb.sjsu.edu/mtiportal/research/publications/summary/a0801.html

Zia, A. and M.H. Glantz. 2012. "Risk zones: comparative lesson drawing and policy learning from flood insurance programs." *Journal of Comparative Policy Analysis: Research and Practice* 14(2): 143–159.

Zia, A. and C. Koliba. 2011. "Accountable climate governance: dilemmas of performance management across complex governance networks." *Journal of Comparative Policy Analysis: Research and Practice* 13(5): 479–497.

Zia, A. and A.M. Todd. 2010. "Evaluating the effects of ideology on public understanding of climate change science: how to improve communication across ideological divides?" *Public Understanding of Science* 19(6): 743–761.

Zia, A., P. Hirsch, A. Songorwa, D.R. Mutekanga, S. O'Connor, T. McShane, P. Brosius, and B. Norton. 2011. "Cross-scale value trade-offs in managing social-ecological systems: the politics of scale in Ruaha National Park, Tanzania." *Ecology and Society* 16(4): 7.

Ziervogel, G. and A. Taylor. 2008. "Feeling stressed: integrating climate adaptation with other priorities in South Africa." *Environment: Science and Policy for Sustainable Development* 50(2): 32–41.

Index

Page numbers in *italics* denote tables, those in **bold** denote figures.